# ZIML Math Competition Book

## Division M 2016-2017

## Areteem Institute

Edited by    John Lensmire
             David Reynoso
             Kevin Wang
             Kelly Ren

Cover and chapter title photographs by Kelly Ren and Kevin
Wang

PUBLISHED BY ARETEEM PUBLISHING
WWW.ARETEEM.ORG
ALL RIGHTS RESERVED.

ISBN: 1-944863-11-7
ISBN-13: 978-1-944863-11-1

First printing, March 2018.

# Contents

# Introduction

Each month during the school year, Areteem Institute hosts the online Zoom International Math League (ZIML) competitions. Students can compete in one of five divisions based on their age and mathematical level.

The book contains the problems, answers, and full solutions from the nine ZIML Division M Competitions held during the 2016-2017 School Year. It is divided into three parts:

1. The complete Division M ZIML Competitions (20 questions per competition) from October 2016 to June 2017.
2. The solutions for each of the competitions, including detailed work and helpful tricks.
3. An appendix including the topics and knowledge points covered for Division M, a glossary including common mathematical terms, and answer keys for each of the competitions so students can easily check their work.

The questions found on the ZIML competitions are meant to test your problem solving skills and train you to apply the knowledge you know to many different applications. We hope you enjoy the problems!

## About Zoom International Math League

The Zoom International Math League (ZIML) has a simple goal: provide a platform for students to build and share their passion for math and other STEM fields with students from around the globe. Started in 2008 as the Southern California Mathematical Olympiad, ZIML has a rich history of past participants who have advanced to top tier colleges and prestigious math competitions, including American Math Competitions, MATHCOUNTS, and the International Math Olympaid.

The ZIML Core Online Programs, most available with a free account at ziml.areteem.org, include:

- **Daily Magic Spells:** Provides a problem a day (Monday through Friday) for students to practice, with full solutions available the next day.
- **Weekly Brain Potions:** Provides one problem per week posted in the online discussion forum at ziml.areteem.org. Usually the problem does not have a simple answer, and students can join the discussion to share their thoughts regarding the scenarios described in the problem, explore the math concepts behind the problem, give solutions, and also ask further questions.
- **Monthly Contests:** The ZIML Monthly Contests are held the first weekend of each month during the school year (October through June). Students can compete in one of 5 divisions to test their knowledge and determine their strengths and weaknesses, with winners announced after the competition.
- **Math Competition Practice:** The Practice page contains sample ZIML contests and an archive of AMC-series tests for online practice. The practices simulate the real contest environment with time-limits of the contests automatically controlled by the server.
- **Online Discussion Forum:** The Online Discussion Forum

is open for any comments and questions. Other discussions, such as hard Daily Magic Spells or the Weekly Brain Potions are also posted here.

These programs encourage students to participate consistently, so they can track their progress and improvement each year.

In addition to the online programs, ZIML also hosts onsite Local Tournaments and Workshops in various locations in the United States. Each summer, there are onsite ZIML Competitions at held at Areteem Summer Programs, including the National ZIML Convention, which is a two day convention with one day of workshops and one day of competition.

ZIML Monthly Contests are organized into five divisions ranging from upper elementary school to advanced material based on high school math.

- **Varsity:** This is the top division. It covers material on the level of the last 10 questions on the AMC 12 and AIME level. This division is open to all age levels.
- **Junior Varsity:** This is the second highest competition division. It covers material at the AMC 10/12 level and State and National MathCounts level. This division is open to all age levels.
- **Division H:** This division focuses on material from a standard high school curriculum. It covers topics up to and including pre-calculus. This division will serve as excellent practice for students preparing for the math portions of the SAT or ACT. This division is open to all age levels.
- **Division M:** This division focuses on problem solving using math concepts from a standard middle school math curriculum. It covers material at the level of AMC 8 and School or Chapter MathCounts. This division is open to all students who have not started grade 9.

- **Division E:** This division focuses on advanced problem solving with mathematical concepts from upper elementary school. It covers material at a level comparable to MOEMS Division E. This division is open to all students who have not started grade 6.

This problem book features the Division M Contests. For a detailed list of topics covered for Division M see p.153 in the Appendix.

## About Areteem Institute

Areteem Institute is an educational institution that develops and provides in-depth and advanced math and science programs for K-12 (Elementary School, Middle School, and High School) students and teachers. Areteem programs are accredited supplementary programs by the Western Association of Schools and Colleges (WASC). Students may attend the Areteem Institute through these options:

- Live and real-time face-to-face online classes with audio, video, interactive online whiteboard, and text chatting capabilities;
- Self-paced classes by watching the recordings of the live classes;
- Short video courses for trending math, science, technology, engineering, English, and social studies topics;
- Summer Intensive Camps on prestigious university campuses and Winter Boot Camps;
- Practice with selected daily problems for free, and monthly ZIML competitions at ziml.areteem.org.

The Areteem courses are designed and developed by educational experts and industry professionals to bring real world applications into STEM education. The programs are ideal for students who wish to build their mathematical strength in order to excel academically and eventually win in Math Competitions (AMC, AIME, USAMO, IMO, ARML, MathCounts, Math Olympiad, ZIML, and other math leagues and tournaments, etc.), Science Fairs (County Science Fairs, State Science Fairs, national programs like Intel Science and Engineering Fair, etc.) and Science Olympiad, or purely want to enrich their academic lives by taking more challenges and developing outstanding analytical, logical thinking and creative problem solving skills.

Since 2004 Areteem Institute has been teaching with methodology that is highly promoted by the new Common Core State Standards: stressing the conceptual level understanding of the math concepts, problem solving techniques, and solving problems with real world applications. With the guidance from experienced and passionate professors, students are motivated to explore concepts deeper by identifying an interesting problem, researching it, analyzing it, and using a critical thinking approach to come up with multiple solutions.

Thousands of math students who have been trained at Areteem achieved top honors and earned top awards in major national and international math competitions, including Gold Medalists in the International Math Olympiad (IMO), top winners and qualifiers at the USA Math Olympiad (USAMO/JMO), and AIME, top winners at the Zoom International Math League (ZIML), and top winners at the MathCounts National. Many Areteem Alumni have graduated from high school and gone on to enter their dream colleges such as MIT, Cal Tech, Harvard, Stanford, Yale, Princeton, U Penn, Harvey Mudd College, UC Berkeley, UCLA, etc. Those who have graduated from colleges are now playing important roles in their fields of endeavor.

Further information about Areteem Institute, as well as updates and errata of this book, can be found online at http://www. areteem.org.

## Acknowledgments

This book contains the Online ZIML Division M Problems from the 2016-17 school year. These problems were created and compiled by the staff of Areteem Institute. These problems were inspired by questions from the Areteem Math Challenge Courses, past questions on the ACT/SAT/GRE, past math competitions, math textbooks, and countless other resources and people encountered by the Areteem Curriculum Department in their life devoted to math. We thank all these sources for growing and nurturing our passion for math.

The Areteem staff, including John Lensmire, David Reynoso, Kevin Wang, and Kelly Ren, are the main contributors who compiled, edited, and reviewed this book. Photographs included on the cover and chapter introduction pages are credit to Kelly Ren and Kevin Wang.

Lastly, thanks to all the students who have participated and continue to participate in the Zoom International Math League. Your dedication to the Daily Magic Spells and Monthly Contests makes all of this possible, and we hope you continue to enjoy ZIML for years to come!

# 1. ZIML Contests

This part of the book contains the Division M ZIML Contests from the 2016-17 School Year. There were nine monthly competitions, held on the dates found below:

- October 7-8
- November 4-6
- December 2-4
- January 6-8
- February 3-5
- March 3-5
- April 7-9
- May 5-7
- June 2-4

## 1.1  ZIML October 2016 Division M

Below are the 20 Problems from the Division M ZIML Competition held in October 2016.
The answer key is available on p.163 in the Appendix.
Full solutions to these questions are available starting on p.74.

### Problem 1

Suppose 100 students at a school were polled about whether they liked math and/or science. 60 of the students liked science, and 50 of the students liked math AND science. 30 students liked neither math nor science. How many of the students like only math?

### Problem 2

Consider the two squares with side length 10 attached together as in the diagram below.

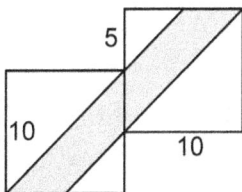

Find the area of the shaded region.

### Problem 3

Consider the number 30800. $a$ and $b$ are the two smallest prime numbers that are not factors of 30800, and $a$ is less than $b$. What is $b$?

## Problem 4

On a good day, Chris the Squirrel picks 20 hazelnuts. On a rainy day he only picks 12 hazelnuts. During a few consecutive days he picked a total of 120 hazelnuts with an average of 15 per day. How many days were rainy?

## Problem 5

Six friends line up to take a photograph. How many different photos are there if two of the friends are a couple who want to stand together?

## Problem 6

Bill usually takes 50 minutes to groom the horses. After working for 10 minutes, he was joined by Ann and they finished the grooming in 15 minutes. How long would it have taken Ann working alone? Express your answer in minutes.

## Problem 7

A number $\overline{35a2a}$ is divisible by 3. Its last three digits form a three-digit number $\overline{a2a}$ that can be exactly divided by 2. Find this five-digit number.

## Problem 8

Two rectangles and two squares are assembled to form a big square as shown.

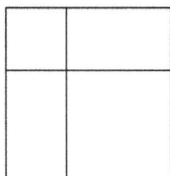

The area of each rectangle is 28 and the area of the small square is 16. What is the area of the entire square?

## Problem 9

Suppose there is a club of 15 students. They vote to elect a president, vice-president, and treasurer. Each student votes for a different person for each office and cannot vote for themselves. How many ways can one student vote?

## Problem 10

In the multiplication problem below, $A$, $B$, $C$ and $D$ are distinct digits. What is $A + B$?

$$
\begin{array}{ccccc}
 & A & B & A & B \\
\times & & & C & D \\
\hline
 & C & D & C & D & B \\
\end{array}
$$

## Problem 11
Judy is playing basketball. After Judy takes 20 shots, she has made 55% of her shots. She takes 5 more shots and she raises her percentage to 56%. How many of the last 5 shots did she make?

## Problem 12
In the diagram, the largest circle has radius 5, and the two smaller circles have radii 3 and 4 respectively. The region $A$ is the over-lapped region of the two smaller circles.

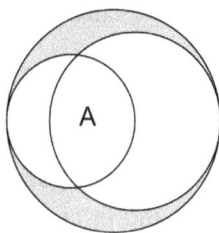

Find the difference between the area of the shaded region and the area of region $A$.

## Problem 13
Adam takes a train to go visit a friend who lives in a city that is 360 kilometers away. The train left his home station at 8:35 AM, and arrived at the destination station at 1:05 PM. How fast has the train traveled measured by average speed in km/h?

## Problem 14
Suppose you flip a coin 7 times. How many outcomes have more heads than tails?

## Problem 15

A jar contains 99 jelly beans. Some students pass the jar around clockwise and eat a jelly bean every time they get the jar. Suppose the 8th friend is the last one to get a jelly bean. Then there are either $G$ friends or $H$ friends, where $G < H$. What is $H - G$?

## Problem 16

Suppose in the diagram below that $\triangle ABC$ is isosceles and $\angle CAG = 20°$.

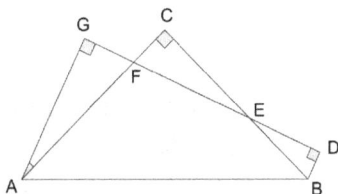

Find the measure of $\angle ABD$ in degrees.

## Problem 17

Brian and David run along a circular track, starting from the same point, going opposite directions. They meet after 36 seconds. Assume that David runs the whole circle in 90 seconds. How long does it take Brian to run the whole circle in seconds?

## Problem 18

Suppose Bill has a dart board with radius 3 feet. Whenever he throws a dart, it randomly lands somewhere on the board. The probability that Bill's dart lands between 1 and 2 feet from the center is $\dfrac{x}{9}$ where $x$ is an integer. What is $x$?

## Problem 19

What is the units digit of $18672^{955}$?

## Problem 20

A sheep is tied at the upper-left corner of the square barn on the grass field. The length and width of the barn are 10, as shown in the diagram, and the length of the rope is 20.

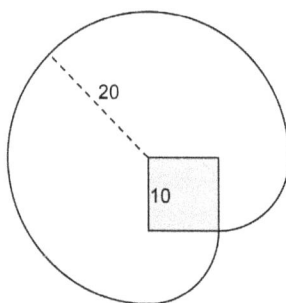

The area of the region that the sheep can reach is $\pi \times A$ for a number $A$. What is $A$?

## 1.2 ZIML November 2016 Division M

Below are the 20 Problems from the Division M ZIML Competition held in November 2016.
The answer key is available on p.164 in the Appendix.
Full solutions to these questions are available starting on p.79.

### Problem 1
How many angles less than $180°$ are there in the following diagram?

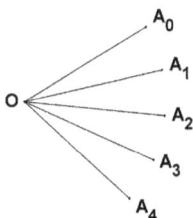

### Problem 2
The ratio of llamas to ostriches in the Areteem petting zoo is $4:7$. If there are a total of 44 llamas and ostriches in the petting zoo, how many of the them are llamas?

### Problem 3
Suppose you are given the list of numbers $1, 4, 6, 4, 3, 8, 2$. What (single) number do you need to insert into the list so that the average (also called the mean) changes to 5?

## Problem 4

Among the fractions

$$\frac{1}{24}, \frac{2}{24}, \frac{3}{24}, \ldots, \frac{23}{24},$$

how many are irreducible? (Irreducible means the fraction cannot be simplified.)

## Problem 5

A restaurant has 5 choices of appetizers, 3 choices for desserts, and as entrées they offer 7 types of burgers and 4 types of salads. How many different meals (appetizer, entrée, dessert) are available?

## Problem 6

Consider the diagram below:

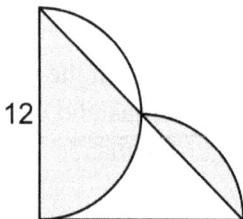

The area of the shaded region is $D \times \pi$ where $D$ is an integer. What is $D$?

## Problem 7

Joe's mom can clean the kitchen in 45 minutes. If Joe helps his mother, they can clean it in 30 minutes. How many minutes would it take Joe to clean it himself?

## Problem 8

A group of 68 people rent 24 motorcycles of two kinds at a racetrack. The first kind has a capacity of 2 people and costs $40 per motorcycle. The second has a capacity of 3 people and costs $30 per motorcycle. The 68 people exactly fill all vehicles. What is the total cost in renting the 24 motorcycles in dollars?

## Problem 9

Suppose we have a star diagram as below (do not assume it is drawn to scale).

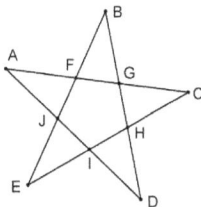

If $\angle A = 40°$ and $AF = AJ$, calculate $\angle B + \angle D$ in degrees.

## Problem 10

Suppose you roll two dice and you only care about what you rolled (not which die has which number). For example, you could roll two 1's or you could roll one 5 and one 6. How many different outcomes are possible?

## Problem 11

How many numbers between 99 and 500 have an odd number of factors?

## Problem 12

Suppose we have a solid ball with radius 4. Suppose you cut the ball in half. The surface area of the half-ball is $K \times \pi$ where $K$ is an integer. What is $K$?

## Problem 13

Katie went hiking on a hill near her home. From the bottom of the hill, She went up to the top and then came down along the same trail, back to the spot she started. Assume her uphill speed was 3 miles per hour, and her downhill speed was 6 miles per hour. What is her average speed for the whole uphill-downhill trip in miles per hour?

## Problem 14

Find the units digit of $2^{2016}$.

## Problem 15

Suppose you flip a fair coin 6 times. The probability that there are no two heads in a row and no two tails in a row is $\dfrac{N}{M}$ where the fraction $\dfrac{N}{M}$ is fully simplified. What is $N+M$?

## Problem 16

The parallelogram $ABCD$ has area 400 cm$^2$ and $E$ is the midpoint of $\overline{AD}$. Find the area of the shaded region.

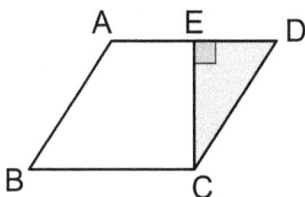

## Problem 17

At 6 AM, bus station A starts to dispatch buses to station B, and station B starts to dispatch buses to station A. They each dispatch one bus to the other station every 8 minutes. The one-way trip takes 45 minutes. One passenger gets on the bus at station A at 6:16 AM. How many buses coming from station B will the passenger see en route?

## Problem 18

Suppose you pay $5 to play a game at the carnival. The game is simple, you press a button and a dial randomly picks a number on a number line between 2 and 14. You then win the amount of money shown (in dollars). What percentage of the time will you win at least as much as you paid to play the game? Input your answer rounded to the nearest percent. For example 82.5% should be submitted as 83.

## Problem 19

Say a whole number is a prime-prime number if it is a prime number that yields a prime number when its units digit is removed. For example, 131 is a three-digit prime-prime number because 131 is prime and 13 is prime. How many two-digit prime-primes are there?

## Problem 20

When several thieves tried to divide a sum of money by giving 8 dollars to each thief, 5 thieves received nothing. When each thief took 7 dollars, they had 5 dollars left over. How many thieves are there in total?

## .3  ZIML December 2016 Division M

Below are the 20 Problems from the Division M ZIML Competition held in December 2016.
The answer key is available on p.165 in the Appendix.
Full solutions to these questions are available starting on p.87.

### Problem 1
Henry reads 160 pages of a book per day. After 5 days, Henry has $\frac{3}{5}$ of the book remaining. How many pages does the book have?

### Problem 2
A big rectangle is divided into 10 smaller congruent rectangles, as shown. Given that the area of the big rectangle is 120 cm², find the perimeter of the big rectangle in cm

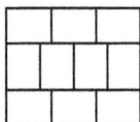

### Problem 3
A merchant offers a large group of items at 30% off. Later, the merchant takes 20% off these sale prices and claims that the final price of these items is 50% off the original price. The true total discount is $K\%$. What is $K$?

## Problem 4

A pool has two inlet pumps A and B. If pump A alone is open, it takes 12 hours to fill the pool with water. If pump B alone is open, it takes 18 hours to fill the pool with water. If the pool needs to be filled in 10 hours, what is the least amount of time (in hours) both pumps need to be open?

## Problem 5

Find the smallest 4-digit number that is divisible by 5 and by 11.

## Problem 6

Given the figure below, if $\angle DFE = 75°$ and $\angle BCF = 95°$, what is the measure of $\angle CAF$?

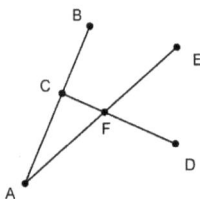

## Problem 7

There are 100 birds and cats. In total there are 80 more bird legs than cat legs. How many birds are there?

## Problem 8

Find the area of the shaded region in the diagram below.

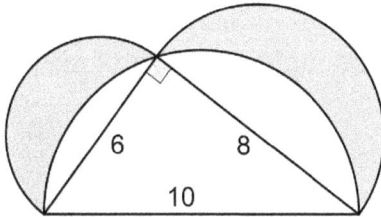

## Problem 9

Consider three cities: City A, City B, and City C. There are 5 one-way roads from $A \to B$, 4 one-way roads from $B \to A$, 3 one-way roads from $B \to C$, and 2 one-way roads from $C \to A$ (and no other roads connect the cities). How many loops are there from City A back to City A (without visiting any other city twice)? Note a loop must leave City A, but does not need to visit all the other cities.

## Problem 10

At the end of the year my school has a talent show. Each participant has exactly 10 minutes to do their presentation. Mr. Tahoe arrived late, and only stayed for 30 minutes. What is the maximum number of participants Mr. Tahoe would be able to see on stage?

## Problem 11

How many factors of 4000 are perfect squares? Remember that 1 and 4000 are both factors of 4000.

## Problem 12

Suppose we are given a bag of 30 balls and 20 cubes. It is known that of the 50 objects in the bag, 40 of them are red and 10 of them are blue. If at least one of the objects in the bag is a blue cube, what is the minimum number of red balls in the bag?

## Problem 13

Concatenate the positive integers $1, 2, 3, \ldots, 2016$ to form a new integer:

$$12345678910111213 14 \cdots 201420152016.$$

What is the remainder when this new integer is divided by 9?

## Problem 14

Consider the shortest path from $A$ to $B$ along the surface of the $2 \times 2 \times 2$ cube shown below.

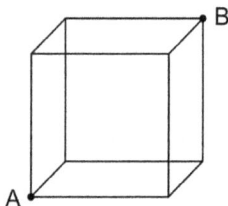

The length of this path is $\sqrt{K}$ for an integer $K$. What is $K$?

## Problem 15

Consider a deck of cards that has 10 blue cards, 10 green cards, 10 yellow cards, and 10 red cards, where the cards of each color are numbered 1 through 10. You are dealt two of the cards randomly. The probability that the second card you are dealt is the same color as the first card is $\dfrac{N}{M}$ where the fraction is in lowest terms. What is $N + M$?

## Problem 16

In the diagram below, there are 21 grid points arranged in equilateral triangles, equally spaced. The AREA of each small equilateral triangle formed by 3 adjacent grid points is 1. Find the area of $\triangle ABC$.

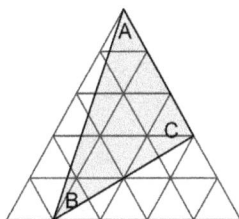

## Problem 17

Sam walks up a hill. After every 30 minutes of walking he takes 10 minutes to rest. When he walks down the hill, he instead rests for 5 minutes after every 30 minutes of walking. Sam walks downhill 1.5 times faster than he walks uphill. If he spends 3 hours and 50 minutes traveling up the hill (including rest), how many minutes does Sam spend traveling down the hill (including rest)?

## Problem 18

Suppose you know a list contains 5 integers. Each number is one of $1, 2, 3, 4, \ldots, 10$ and it is possible that some numbers are repeated. How many different possibilities are there for the average of the list of 5 numbers?

## Problem 19

The least common multiple of two numbers is 180 and their greatest common divisor is 30. One of the two numbers is 90. What is the other number?

## Problem 20

What is the next term in the sequence $1, 2, 9, 28, 65, 126, \_\_$.

## 1.4 ZIML January 2017 Division M

Below are the 20 Problems from the Division M ZIML Competition held in January 2017.
The answer key is available on p.166 in the Appendix.
Full solutions to these questions are available starting on p.96.

### Problem 1

A certain fraction, $\frac{m}{n}$, if reduced to lowest terms, equals $\frac{5}{11}$. Also given that $m + n = 80$. What is $m$?

### Problem 2

Two rectangles and one square are assembled to form a big square as shown. The areas of the rectangles are 44 and 28. What is the area of the smaller (lower-right) square?

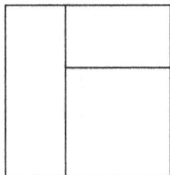

### Problem 3

Find the smallest positive integer consisting only of the digits 3 and 5, with at least one of each, that is divisible by 3 and 5.

## Problem 4

George wrote a computer program. If you type in a positive integer, George's program outputs the sum of all the positive factors of that integer. For example, if you type in 6, the program outputs 12, because the factors of 6 are $1, 2, 3, 6$ and $1 + 2 + 3 + 6 = 12$. Suppose George's friend Paul types in 7 and then runs the program. If Paul types in this result and runs the program again, what is the output?

## Problem 5

It requires either 45 small trucks or 36 big trucks to transport a batch of steel blocks. Given that each big truck can load 4 more tons of steel blocks than each small truck, how many tons of steel blocks are in a batch?

## Problem 6

Suppose that a restaurant sells 7 different burgers and 4 different salads. Two people both order a burger or both order a salad, but not the same burger or the same salad. How many different ways can this happen?

## Problem 7

Bella goes shopping at the marketplace for shawls and belts. The shawls she likes each cost $12. The belts she likes each cost $14. Bella has exactly enough money to buy a certain number of shawls. If she buys belts instead, she has exactly enough money to buy 3 fewer belts. How much money in dollars did Bella bring with her to the market?

## Problem 8

In the diagram, the radius of the circle is 4. If we are also given that the area of the shaded region is $14\pi$, find the area of the triangle.

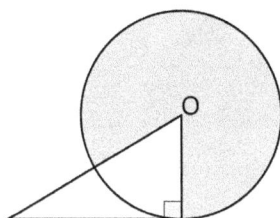

Note: This diagram does not reflect the actual shape: the angle of the unshaded sector is not $60°$.

## Problem 9

It takes $\frac{1}{3}$ more time for Andy to plant one tree than for Nathan. If Andy and Nathan work together, then in the end Nathan plants 36 more trees than Andy does. How many trees are in total?

## Problem 10

Suppose you have 5 red and 8 green balls. You pick 5 balls at once (in no particular order and without replacing the balls). The probability that all 5 balls are the same color is $\frac{P}{Q}$ where the fraction is written in lowest terms. What is $Q - P$?

## Problem 11

Suppose five squares are attached to a regular pentagon as in the diagram below.

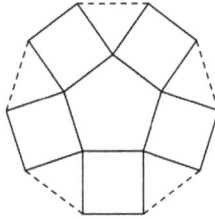

Consider the 5 (congruent) triangles formed by connecting vertices of the squares. What is the difference between the largest and smallest angles of these triangles (measured in degrees)?

## Problem 12

As Emily is riding her bicycle on a long straight road, she spots Emerson skating in the same direction $1/2$ mile in front of her. After she passes him, she can see him in her rear view mirror until he is $1/2$ mile behind her. Emily rides at a constant rate of 12 miles per hour, and Emerson skates at a constant rate of 8 miles per hour. For how many minutes can Emily see Emerson?

## Problem 13

Find the distance from one corner to the opposite corner (for example the lower front left vertex to the upper back right vertex) of a rectangular prism with dimensions 3, 4, and 12. That is, find the length of the line connecting the two corners. Round your answer to the nearest integer if necessary.

## Problem 14
How many 4 digit numbers that do not contain the digit 0 are there that have digits that sum up to 8?

## Problem 15
The SAT is a renowned high school standardized test administered for college admissions purposes. The scores are in multiples of 10 (maximum 800 per section) and the total score is determined by adding the critical reading subscore and the math subscore. Of the 50 students that scored above 600 on either the critical reading or math section of the SAT, 35 students scored above 600 on the math section and 25 students scored above 600 on the critical reading section. What is the maximum average total score of all 50 students?

## Problem 16
The table below gives the percent of students in each grade at school A and school B.

|   | K | 1 | 2 | 3 | 4 | 5 | 6 |
|---|---|---|---|---|---|---|---|
| A | 21% | 12% | 11% | 15% | 13% | 17% | 11% |
| B | 18% | 11% | 16% | 11% | 13% | 14% | 17% |

School A has 100 students, and school B has 200 students. If the two schools combined, what percent of the students are in grade 6?

## Problem 17
The sum of the reciprocals of four consecutive primes is $\dfrac{N}{1155}$. What is $N$?

## Problem 18

All the smaller circles in the diagram below have radii 1.

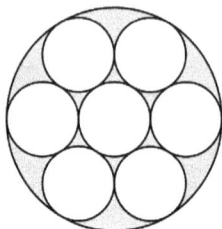

The combined perimeter of all the shaded regions is $L \times \pi$ where $L$ is an integer. What is $L$?

## Problem 19

Suppose you randomly pick a number/point on the number line between $-3$ and 7. (So the number could be 2, $-1.6$, $\pi$, $\sqrt{2}$, etc.) The probability that the number squared is greater than 4 is approximately $K\%$ where $K$ is an integer. What is $K$?

## Problem 20

Three numbers, 22, 41, and 60, are divided by a positive integer $d$, and the three remainders are $r_1$, $r_2$, and $r_3$, respectively. Given that $r_1 + r_2 + r_3 = 21$, determine the number $d$.

## .5  ZIML February 2017 Division M

Below are the 20 Problems from the Division M ZIML Competition held in February 2017.
The answer key is available on p.167 in the Appendix.
Full solutions to these questions are available starting on p.107.

### Problem 1
A goldsmith has 10 grams of a 40% gold alloy (an alloy is a mixture of metals). How many grams of pure gold should be added to make an alloy which is 60% gold?

### Problem 2
It takes 25 seconds for a train to pass completely through a tunnel which measures 250 meters long. It takes 23 seconds for the train to completely pass through another tunnel which measures 210 meters long. How many seconds does it take the train to pass an approaching train which is 320 meters long and traveling at a speed of 18 m/s?

### Problem 3
Two trucks dump 400 cubic meters of dirt. Truck A carries 7 cubic meters per load. Truck B carries 4 cubic meters per load. The dirt is removed after 70 loads. How many loads are carried by truck A?

## Problem 4

There are 2 packs of crayons available for every 5 students at Amy's art class. How many students can share 18 packs of crayons?

## Problem 5

Alice, Bob, and Cindy drive their cars separately from site A to site B simultaneously. Alice drives at 60 mph and Bob drives at 48 mph. Alice passes a car from the opposite direction after 6 hours of driving. One hour later, Bob pass the same car still traveling in the opposite direction. One more hour later, Cindy also passes the same car. Find the speed at which Cindy drives her car. Give your answer in mph.

## Problem 6

Find the side length of a regular hexagon with area $150\sqrt{3}$.

## Problem 7

Given square $ABCD$, let $P$ and $Q$ be the points outside the square that make triangles $CDP$ and $BCQ$ equilateral. Segments $\overline{AQ}$ and $\overline{BP}$ intersect at $G$. Find angle $AGP$ in degrees.

## Problem 8

Suppose you have a right triangle $\triangle ABC$ with $\angle B = 90°$, $AB = 8$ and $BC = 6$. Extend the line segment $BC$ to point $D$ so that the obtuse triangle $\triangle ACD$ is formed. If $CD = 9$, find the perimeter of $\triangle ACD$.

## Problem 9

In rectangle $ABCD$, let $O$ be a point in the rectangle. If the area of $\triangle AOB$ is 6 cm$^2$, and the area of $\triangle DOC$ is 12 cm$^2$, find the area of rectangle $ABCD$.

## Problem 10

Parallelogram $ABCD$ is shown below, where triangle $BCE$ is a right isosceles triangle and $A$ is the midpoint of $\overline{BE}$.

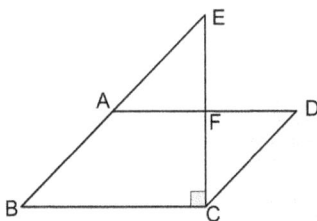

Given that $[ABCF] - [DFC] = 4$, find the area of $ABCD$.

## Problem 11

Alice and Bob are interested in furnishing their apartment. They wanted to buy a couch, a coffee table, and a recliner. After an intensive search for furniture, they have narrowed it down to 3 couches, 6 coffee tables, and 5 recliners. How many different furniture combinations are possible?

## Problem 12

In the Chocolate Factory, Willy Wonka serves three different types of chocolates: Hershey Kisses, Reese's Peanut Butter Cups, and Kit Kat bars. Willy Wonka decides to be generous and allows you to have exactly 5 pieces of candy for free. How many different ways can you do choose which chocolates to get?

## Problem 13

How many different rearrangements of the word ARETEEM are there?

## Problem 14

6 women and 8 men are on the faculty in the mathematics department at a school. How many ways are there to select a committee of 5 members of the department if at least 2 woman must be on the committee?

## Problem 15

In a game show, you are presented a mystery box containing 5 U.S. dollar bills, but the denomination of each dollar bill is unknown. The host claims that there are denominations of either $1, $10, $100, or $1000 contained in the box. How many possible monetary values are contained in the box?

## Problem 16

A math class with 20 students take an exam. A fourth of the students who passed got an $A$ and a third who passed got a $B$. How many students passed the exam?

## Problem 17

A number $\overline{35a2a}$ is divisible by 2. Its last three digits form a three-digit number $\overline{a2a}$ that can be exactly divided by 3. Find the greatest such five-digit number.

## Problem 18

There are two two-digit numbers whose square ends in the same two-digit number. Find the larger of them.

## Problem 19

What is the units digit of $7^{47} + 7^{74}$?

## Problem 20

What is the greatest prime factor of $13^{234} + 13^{235} + 13^{236}$?

## 1.6  ZIML March 2017 Division M

Below are the 20 Problems from the Division M ZIML Competition held in March 2017.
The answer key is available on p.168 in the Appendix.
Full solutions to these questions are available starting on p.116.

### Problem 1
If the area of the rectangle is 40, the area of the smaller white triangle is 5, and the area of the big white triangle is 20, what is the area of the shaded region?

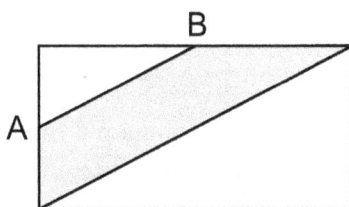

### Problem 2
Suppose you have a soccer tournament consisting of 8 teams. Each team plays every other team twice: once on their home field and once on the opponents field. How many games in total are there in the tournament?

### Problem 3
Rita has 36 marbles, 20 of which are red and 16 of which are white. Rosa also has red and white marbles, 27 in total, in the same proportion as Rita. How many white marbles does Rosa have?

## Problem 4

For this problem, approximate $\pi \approx 3.14$. Using this approximation, the area of shaded region $A$ minus the area of shaded region $B$ in the diagram below is $L$, where $L$ is an integer. What is $L$?

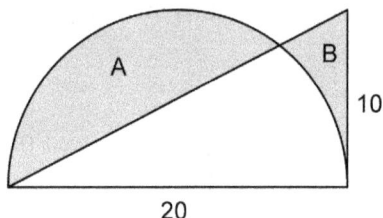

## Problem 5

Suppose 8 people run a race. If the top 4 finishers in the race can advance to the next race, how many different groups can advance?

## Problem 6

Find the largest two-digit number that leaves a remainder of 2 when divided by 4 and 6.

## Problem 7

Stephanie begins walking at a pace of 4 km per hour from one end of a trail that is 50 km long. Bob begins at the other end of the trail at the same time, walking towards Stephanie at a pace of 6 km per hour. How many hours will it take for them to pass each other?

## Problem 8
How many positive two-digit integers are factors of both 2016 and 2160?

## Problem 9
The star diagram below is not drawn to scale.

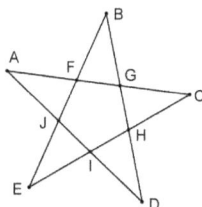

If $\angle A = 30°$ and $\triangle AFJ$ is isosceles with $AF = AJ$, what is $\angle B + \angle D$? Give your answer in degrees.

## Problem 10
A prime-prime number is a prime number that yields a prime number when its units digit is omitted. For example, 131 is a three-digit prime-prime number because 131 is prime and 13 is prime. How many prime-prime numbers are between 200 and 250?

## Problem 11
How many six-digit numbers represented by 2017__ __ are divisible by 3 and 5?

## Problem 12

Suppose a $8 \times 8$ checkerboard is colored red and black. How many ways are there to put a red checker and a black checker on the board so that the checkers are in two squares of different colors?

## Problem 13

If I drive from Irvine to Fullerton at 60 miles per hour and then from Fullerton to Irvine at 40 miles per hour, what is my average speed for the whole journey? Give your answer in miles per hour, rounded to the nearest integer if necessary.

## Problem 14

The price of a pack of colored pencils is $19 and the price of a pack of regular pencils is $11. The math teacher bought 16 packs of pencils for a total of $280. How many packs of colored pencils did he buy?

## Problem 15

Find the distance from one corner to the opposite corner (for example the lower front left vertex to the upper back right vertex) of a rectangular prism with dimensions $8, 9, 12$. That is, find the length of the line connecting the two corners. Round your answer to the nearest integer.

## Problem 16

Suppose you have two sticks, one of length 1 and the other of length 2. You randomly break the stick of length 2 into two pieces. The probability you can use the resulting 3 sticks to form a triangle can be expressed as $\frac{p}{q}$ in lowest terms. Find $p+q$.

## Problem 17

How many gallons of a 10% ammonia solution should be mixed with 50 gallons of a 30% ammonia solution to make a 15% ammonia solution?

## Problem 18

Let $ABC$ be the triangle given below:

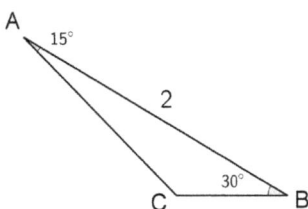

The area of triangle $ABC$ can be expressed in the form $\dfrac{P+\sqrt{Q}}{R}$, where $P$, $Q$ and $R$ are integers and $\sqrt{Q}$ is in simplest radical form. Find $P+Q+R$.

## Problem 19

There is a 4 digit number written on the wall that you know can be divided by 11. However, one of the digits was smudged and you can only see 3__41. What digit is missing?

## Problem 20
How many 8-digit numbers have the sum of their digits being 3?

## .7  ZIML April 2017 Division M

Below are the 20 Problems from the Division M ZIML Competition held in April 2017.
The answer key is available on p.169 in the Appendix.
Full solutions to these questions are available starting on p.123.

### Problem 1
Suppose one number is added to the list 1, 4, 6, 4, 3, 8, 2, so that the average of the new list is 5. What is the number added?

### Problem 2
In the diagram below, there are 36 rectangular grid points, evenly spaced, and the distance between each pair of adjacent points is 1. Find the area of quadrilateral *ABCD*. Express your answer as a decimal, rounded to the nearest hundredth if necessary.

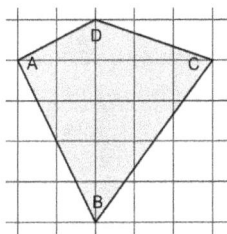

### Problem 3
Jim is paid a 7% commission on the first $700 of his weekly sales, and a 12% commission on any sales past $700. If Jim's sales were $1200, how many dollars was his commission?

## Problem 4

A rectangle is divided into 4 squares, as shown in the diagram.

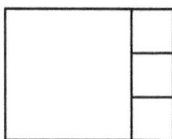

Given that the area of one bigger square is 24 square inches more than the area of one smaller square, find the area of the whole rectangle in square inches.

## Problem 5

Sam is training for a 5 km run. While training he runs for 6 minutes, then rests for 2 minutes, runs for 6 minutes, rests for 2 minutes, etc., until he finishes the 5 km. It takes him 26 minutes in total (with the rest periods) to finish the 5 km rum. If Sam runs at a constant speed (when he is not resting), what is this speed in kilometers per hour? (Recall there are 60 minutes in 1 hour.)

## Problem 6

If Emily and Julia work together, they can finish a project in 15 days. Working separately, it takes the same amount of time for Emily to complete $\frac{1}{2}$ of the project as it takes for Julia to complete $\frac{1}{3}$ of the project. If Emily works alone, how many days does it take for her to complete the project?

## Problem 7

What is the units digit of $14^{200} + 16^{300}$?

## Problem 8

Consider the numbers $1, 2, 3, 4, \ldots, 30$. How many of these numbers have at least 4 factors?

## Problem 9

A number $\overline{12a3a}$ is divisible by 3 (that is, the 5 digit number with digits 1, 2, $a$, 3, and $a$). Its last two digits form a number $\overline{3a}$ that can be exactly divided by 4. Find this five-digit number.

## Problem 10

Consider the shaded region in the diagram below, formed from a right triangle and a sector of a circle.

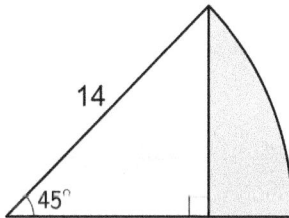

Using the approximation $\pi = \dfrac{22}{7}$, the area of the shaded region is an integer $L$. What is $L$?

## Problem 11
A group of 68 people rent 24 motorcycles of two kinds at a racetrack. The first kind has a capacity of 2 people and costs $40 per motorcycle. The second has a capacity of 3 people and costs $30 per motorcycle. The 68 people exactly fill all vehicles. What is the total cost of renting the 24 motorcycles in dollars?

## Problem 12
Steve has 6 different bars of candy. He wants to bring the candy to eat during lunch the next three days (2 candy bars each day). How many different ways can Steven divide his candy bars into 3 groups to bring to lunch?

## Problem 13
Consider two triangles drawn in the rectangle shown below.

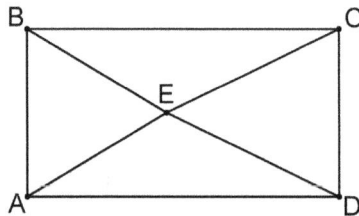

Suppose $\triangle ABE$ is equilateral and that $\angle CDE = 65°$. What is the measure of $\angle AED$ in degrees?

## Problem 14

In a Math Challenge class, the number of students is between 10 and 20. These students sit around a circular table, and start counting off numbers, clockwise, beginning with 1, and continue until 100 is counted. If the numbers 5 and 100 are counted by the same student, how many students are in the class?

## Problem 15

Jane's family loves to come visit her. Her parents come visit every 6 weeks, and her grandparents come visit every 8 weeks. Both Jane's parents and grandparents visited during the same week the last week of 2016. How many times will her parents and grandparents visit during the same week in 2017? (Remember one year has 52 weeks.)

## Problem 16

Candace has a model rocket that she likes to launch. The rocket contains a parachute that returns it to the ground after launch. On a calm day, the rocket lands somewhere randomly within 20 meters of the launch site. Candace has enough room in her yard that if the rocket lands within 5 meters of the launch site she does not need to leave her yard to retrieve her rocket. If she launches the rocket on a calm day, the probability that Candace needs to leave her yard to get the rocket can be written as a reduced fraction $\frac{P}{Q}$. What is $P + Q$?

### Problem 17

Let $A$ be the set of perfect squares in the set of whole numbers from 1 to 100 (inclusive) and let $B$ be the set of perfect cubes in the set of whole numbers from 1 to 100 (inclusive). For example, 1 is in $A$ because $1 = 1^2$ and 1 is also in $B$ because $1 = 1^3$. Find the sum of all the elements that are in $B$ but NOT in $A$.

### Problem 18

How many ways can you place 20 identical balls into 4 distinguishable bins such that there is at least one ball in each bin?

### Problem 19

A railroad bridge measures 1000 meters long. A train passes the bridge. It takes 120 seconds from the time the front of the train train enters the bridge to the time the back of the train train fully leaves the bridge. There are a total of 80 seconds that the entire train is on the bridge. Find the length of the train in meters.

### Problem 20

Billy's first and favorite ball has a radius of 4 inches. For his birthday, he gets a second and third ball. The volume of the second ball is 27 times the volume of his first ball, and the surface area of the third ball is $\frac{1}{16}$ times the surface area of his first ball. Billy balances all three balls (very carefully) in a vertical stack. How high is the stack of three balls? Give your answer in inches.

## .8  ZIML May 2017 Division M

Below are the 20 Problems from the Division M ZIML Competition held in May 2017.
The answer key is available on p.170 in the Appendix.
Full solutions to these questions are available starting on p.131.

### Problem 1
Arrange several equilateral triangles and rhombi, all of whose side lengths are 2 cm, to form a long parallelogram, as shown in the diagram.

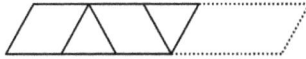

If the perimeter of the long parallelogram is 248 cm, how many rhombi are there?

### Problem 2
There is only one five-digit number that is (i) a multiple of 11, and (ii) made up of only the digits 2 and 3. What is this number?

## Problem 3

You and 6 of your friends are camping at the beach over the weekend. Each of the two nights you will be there you agree to have two shifts of 2 people keeping watch while the others sleep. To make it fair, if someone kept watch on one of the shifts of the night, that person will not keep watch on the other shift of that night. In how many different ways can you decide the 4 shifts to keep watch?

## Problem 4

Four basketballs and five volleyballs cost 185 dollars in total. If a basketball costs 8 dollars more than a volleyball, what is the cost of one basketball in dollars?

## Problem 5

Suppose below is a map of a city where each square in the diagram is a square block.

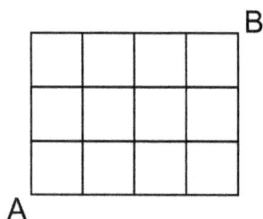

You want to travel from $A$ to $B$ using the least number of blocks possible. How many different routes can you take?

## Problem 6

What are the last two digits of $43^{2015}$?

## Problem 7

Three squares with side lengths 4, 8 and 4 are arranged as in the diagram below.

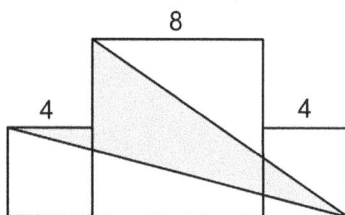

What is the area of the shaded region?

## Problem 8

Carolyn reads a book. Initially, the number of pages that she had read and the number of pages that she had not read were in ratio 3 : 4. After she reads an additional 33 pages of the book, the ratio becomes 5 : 3. How many pages does the whole book have?

## Problem 9

The sum of eleven consecutive even numbers is 374. What was the third number that was added?

## Problem 10

Chris has 18 identical pieces of candy saved to bring to lunch next week. He wants to make sure to bring at least 2 pieces each day, except for Wednesday when he wants to bring at least 4. In how many ways can he sort his candy to bring each of the 5 days of the week to lunch?

## Problem 11
Jong-Zhi took a math test that had 12 arithmetic questions, 15 algebra questions and 18 geometry questions. She got 75% of the arithmetic questions correct and 60% of the algebra questions correct. How many of the geometry questions must she get correct to get at least a passing grade of 75%?

## Problem 12
Suppose in the diagram below that $\triangle ABC$ is isosceles and $\angle CAG = 20°$. Find the measure of $\angle ABD$ in degrees.

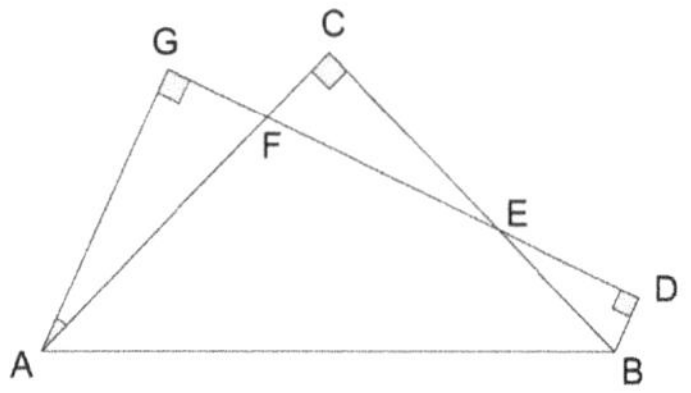

## Problem 13
In a math class, the teacher brings some pencils into the classroom. If he distributed the pencils evenly among the girls, each girl would get 21 pencils. If he distributed the pencils evenly among the boys, each boy would get 28 pencils. In fact, the teacher distributes the pencils evenly among all the students. How many pencils does each student receive?

## Problem 14

Billy and his friends are randomly picking numbers from 1, 2, 3, ..., 10 and adding them to a list. They have picked 9 numbers so far, getting

$$3, 7, 2, 2, 9, 1, 8, 5, 1.$$

The probability that the $10^{\text{th}}$ number chosen will raise the median can be expressed as a reduced fraction $\dfrac{P}{Q}$. What is $P + Q$?

## Problem 15

In the warehouse there were 3 times as many pounds of apples as pounds of bananas at the beginning. Suppose 250 pounds of bananas and 600 pounds of apples were sold every day; a few days later the bananas were sold out and there were 750 pounds of apples left. How many pounds of apples were there originally?

## Problem 16

George has a ball of radius 6 inches and Carl has two balls each with radius 3 inches. The volume of George's ball minus the combined volumes of Carl's two balls can be written as $K \times \pi$ for an integer $K$. What is $K$?

## Problem 17

Suppose you have a stick of length 1 foot. You get a second stick whose length is randomly chosen from the number line between 0 and 3 feet and a third stick whose length is randomly chosen from the number line between 0 and 5 feet. (So these sticks do NOT have to have lengths that are integers!) The probability that the third stick is longer than the other two combined can be expressed as $N\%$. What is $N$ rounded to the nearest integer?

## Problem 18

What is the smallest integer $K$ so that $252 \times K$ is a perfect cube? Recall 125 is a perfect cube because $5^3 = 125$.

## Problem 19

Starting at the same time, a bus and a truck start traveling toward each other. After 18 hours the two vehicles meet. The bus travels at 50 miles per hour. The truck travels at 42 miles per hour, but stops for a 1 hour break after every 3 hours of travel. How many miles are between the two starting locations?

## Problem 20

In the diagram below, $AB$ has length two and the inner circle has radius 1.

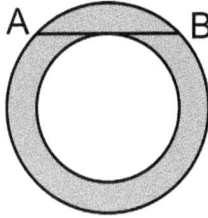

Rounded to the nearest tenth, what is the area of the shaded region?

## 1.9 ZIML June 2017 Division M

Below are the 20 Problems from the Division M ZIML Competition held in June 2017.
The answer key is available on p.171 in the Appendix.
Full solutions to these questions are available starting on p.141.

### Problem 1
Hercules bought 5 swords and 2 spears for 304 drachmas (drachmas were the currency of Greece). A spear cost 9 drachmas less than a sword. How many drachmas does a sword cost?

### Problem 2
Michelle has a dozen oranges and a dozen pears. Assume all the oranges are the same size and all the pears are the same size. Michelle uses her juicer to extract 8 ounces of pear juice from 3 pears and 8 ounces of orange juice from 2 oranges. She makes a pear-orange juice blend from an equal number of pears and oranges. The blend is then $K\%$ pear juice, where $K$ is an integer. What is $K$?

## Problem 3

In the following diagram, $\angle 1 = \angle 2 = \angle 3$.

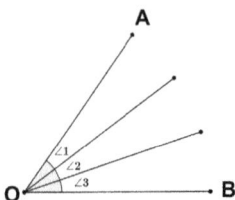

The sum of the measures of all possible acute angles in $\angle AOB$ is $180°$. What is the measure of $\angle AOB$ in degrees?

## Problem 4

A four-digit number $\overline{2A5B}$ should be exactly divisible by 36. Find the sum of all possible four-digit numbers that meet this requirement.

## Problem 5

Suppose we are given a bag of 30 balls and 20 cubes. It is known that of the 50 objects in the bag, 40 of them are red and 10 of them are blue. If at least one of the objects in the bag is a blue cube, what is the minimum number of red balls in the bag?

## Problem 6
Consider the diagram below.

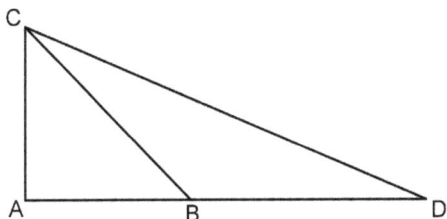

Suppose $AB = AC = 1, \angle BAC = 90°$, and $\angle ADC = 22.5°$. Find $BD^2$ (the length of $BD$ squared).

## Problem 7
Suppose you have a secret. On Sunday you tell it to a friend. On Monday your friend tells your secret to 2 other friends. On Tuesday, each of your friends who heard your secret on Monday tell it to 2 other friends. If this procedure continues, how many people in all (not including you!) will have been told your secret by the end of the following Saturday? Assume no one is told your secret more than once.

## Problem 8
If Iris spends 3 days and Olivia spends 5 days on a project, $\frac{1}{2}$ of the work can be completed. If instead Iris spends 5 days and Olivia spends 3 days on the project, $\frac{1}{3}$ of the work can be completed. How many days would it take for Iris to complete the whole project if she works alone? Round your answer to the nearest day if necessary.

## Problem 9

James, Jim, and 6 other friends line up for a photograph. James and Jim do not stand next to each other. How many different photographs are possible?

## Problem 10

Alice leaves site A toward site B. At the same time Bob leaves site B toward A. Alice drives at 40 miles per hour, and Bob drives at 60 miles per hour. After they pass by each other, Alice drives 4.5 additional hours to arrive at B. How far is it between A and B in miles?

## Problem 11

Two rectangles and one square are assembled to form a big square as shown.

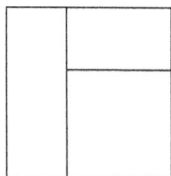

The areas of the rectangles are 44 and 28. What is the area of the smaller (lower-right) square?

## Problem 12

Find the remainder when $2^{2017}$ is divided by 7.

## Problem 13
A school purchases 3 different sizes of projectors, total of 47. The large size costs $700, the medium costs $300, and the small costs $200. The total cost of the projectors is $21200, and there are twice as many medium projectors than small projectors. How many medium projectors does the school purchase?

## Problem 14
How many perfect squares are factors of 308000?

## Problem 15
John, Paul, Ben, and Andy play a game. Paul is twice as likely to win as John is, Ben is three times as likely to win as John is, and Andy is four times as likely to win as John is. Then the probability that Ben wins a game is $L\%$ for an integer $L$. What is $L$?

## Problem 16

Six cylindrical pencils are tied together with a rubber band. A cross section of it is shown in the diagram.

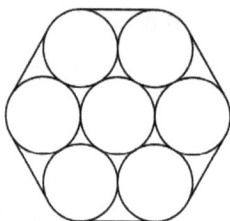

The radius of each pencil is 3 mm. The current length of the rubber band (in mm) can be written in the form $A \times \pi + B$ for integers $A, B$. What is $A + B$?

## Problem 17

Suppose 5 people are to be seated in a row of 10 chairs. How many possible seating arrangements can be made if there must be at least one empty seat between each pair of adjacent people?

## Problem 18

Suppose that when the numbers $513, 571$, and $658$ are all divided by the same integer greater than 1, the remainders are all the same. Find this common remainder.

## Problem 19

Consider $\triangle ABC$ and let $D$ and $G$ be the midpoints of $\overline{AB}$ and $\overline{AC}$, respectively. Further, let $E$ and $F$ be on $\overline{BC}$ such that $BE = EF = FC$. If the area of $\triangle ABC$ is 36 find the area of the quadrilateral $DEFG$. Round your answer to the nearest integer if necessary.

## Problem 20

The DMV is debating a new rule about license plates. Before they allowed only one letter (the 26 letters A, B, C, to Z) followed by 5 digits (the digits 0, 1, up to 9) and the new rule would be to allow 2 letters followed by 4 digits. Letters and digits are allowed to be repeated. How many more license plates can they have with this new rule?

# 2. ZIML Solutions

This part of the book contains the official solutions to the problems from the nine Division M ZIML Contests from the 2016-17 School Year.

Students are encouraged to discuss and share their own methods to the problems using the Discussion Forum on ziml.areteem.org.

## 2.1   ZIML October 2016 Division M

Below are the solutions from the Division M ZIML Competition held in October 2016.

The problems from the contest are available on p.15.

### Problem 1 Solution

30 students do not like math or science, so 70 like math or science (or both). 50 like both, so $60 - 50 = 10$ like only science. This means $70 - 50 - 10 = 10$ like only math.

**Answer:** 10

### Problem 2 Solution

The shaded region is a parallelogram with base $10 - 5 = 5$ and height $10 + 5 = 15$, hence has area $15 \times 5 = 75$.

**Answer:** 75

### Problem 3 Solution

The prime factorization of $30800 = 2^4 \times 5^2 \times 7 \times 11$. Therefore $a = 3$ and $b = 13$.

**Answer:** 13

### Problem 4 Solution

First note that since Chris picked 15 hazelnuts on average and we know that he picked 120 hazelnuts in total, there must have been $120 \div 15 = 8$ days in total. If each of the 8 days were sunny, Chris would have picked $8 \times 20 = 160$ total hazelnuts, $160 - 120 = 40$ more than he actually did. On a rainy day Chris picks $20 - 12 = 8$ fewer hazelnuts than on a sunny day. Therefore, $40 \div 8 = 5$ of the 8 days must have been rainy.

**Answer:** 5

## Problem 5 Solution

There are $2 = 2!$ ways to decide the arrangement of the couple. If we then treat them as a single 'object' we must arrange 5 'objects', which can be done in a total of $5! = 120$ ways. This gives a total of $2 \times 120 = 240$ pictures.

**Answer:** 240

## Problem 6 Solution

Bill can groom $\frac{1}{50}$ of the horses in a single minute. Since Bill works $10 + 15 = 25$ minutes in total, he groomed $25 \times \frac{1}{50} = \frac{1}{2}$ of the horses. Thus, Ann must have groomed the other $\frac{1}{2}$ of the horses in 15 minutes, so it would take her $15 \times 2 = 30$ minutes to groom all the horses working alone.

**Answer:** 30

## Problem 7 Solution

The digit a has to be even. The sum of digits $3 + 5 + a + 2 + a = 10 + 2a$ is divisible by 3. The only digit $a$ that meets this requirement is 4.

**Answer:** 35424

## Problem 8 Solution

The little square's area is 16, so its side length is 4. Thus the small rectangle is $4 \times 7$, and therefore the dimensions of the entire square are $11 \times 11$, hence has area 121.

**Answer:** 121

## Problem 9 Solution

They cannot vote for themselves, so they have 14 choices for president, 13 for vice-president, and 12 for treasurer. In total this

---

gives $14 \times 13 \times 12 = 2184$ possibilities.

**Answer:** $2184$

**Problem 10 Solution**
Note

$$\overline{CDCDB} = 1000\overline{CD} + 10\overline{CD} + \overline{B} = 1010\overline{CD} + \overline{B}.$$

We know $\overline{CDCDB}$ is a multiple of $\overline{CD}$, so $\overline{CD}$ must divide $\overline{B}$, thus $B = 0$. Now it is clear that $A = 1$. Therefore $A + B = 1$.

**Answer:** $1$

**Problem 11 Solution**
Judy made $55\%$ of the first 20 shots, or

$$55\% \times 20 = 0.55 \times 20 = 11$$

shots. If she raises her percentage to $56\%$ after 25 total shots, she must have made a total of

$$56\% \times 25 = 0.56 \times 25 = 14$$

shots. Hence, she made $14 - 11 = 3$ out of the last 5 shots.

**Answer:** $3$

**Problem 12 Solution**
Since $3^2 + 4^2 = 5^2$, the area of the largest circle equals the sum of the two smaller circles. Thus the region A and the shaded region have the same area. Therefore the answer is 0.

**Answer:** $0$

**Problem 13 Solution**
The total time Adam traveled is 4 hours and 30 minutes, which is 4.5 hours. Since the train traveled a total of 360 km, the train has

an average speed $30 \div 4.5 = 80$ km/h so the answer is 80.

**Answer:** 80

## Problem 14 Solution

Note the number of outcomes that have more heads is equal to the number of outcomes with more tails (for any outcome, switch heads and tails to get an outcome of the opposite type). There are $2^7 = 128$ total outcomes, so 64 have more heads than tails.

**Answer:** 64

## Problem 15 Solution

The number of friends must be a factor of $99 - 8 = 91 = 7 \cdot 13$. Since there are more than 7 friends, the answer must be 13 or 91 friends so $G = 13, H = 91$. Hence $H - G = 78$.

**Answer:** 78

## Problem 16 Solution

First note as $\triangle ABC$ is an isosceles right triangle, $\angle ABC = 45°$. Since $\angle CAG = 20°$, $\angle AFG = 70°$. Continuing in this manner, we see $\angle CAG = \angle CEG = \angle DEB = 20°$ and $\angle AFG = \angle CFE = \angle DBE = 70°$. Therefore, $\angle ABD = 70 + 45 = 115°$.

**Answer:** 115

## Problem 17 Solution

Since David runs the whole circle in 90 seconds, at the time they meet (36 seconds), David finishes $\dfrac{36}{90} = \dfrac{2}{5}$ of a whole circle, so Brian finishes $1 - \dfrac{2}{5} = \dfrac{3}{5}$ of a whole circle. Thus the total time it takes for Brian to run the whole circle is $36 \div \dfrac{3}{5} = 60$ seconds.

**Answer:** 60

## Problem 18 Solution

The region we want is inside a circle of radius 2, and outside a circle of radius 1, so has area $\pi 2^2 - \pi 1^2 = 3\pi$. We then divide this by the full circle which has area $\pi 3^2 = 9\pi$. Hence the probability is $3/9$ so $x = 3$.

**Answer:** 3

## Problem 19 Solution

Only look at the last digit: 2. The units digits of powers of 2 have a pattern: $2, 4, 8, 6, 2, 4, 8, 6, \ldots$, which repeats every 4 terms. Since $955 \div 4$ has remainder 3, our answer is the 3rd term in the pattern: 8.

**Answer:** 8

## Problem 20 Solution

The region can be divided into $3/4$ of a circle with radius 20 and two quarter circles with radii 10. Thus the area is $\frac{3}{4} \times 20^2 \pi + 2 \times \frac{1}{4} \times 10^2 \pi = 350\pi$. Hence $A = 350$.

**Answer:** 350

## 2 ZIML November 2016 Division M

Below are the solutions from the Division M ZIML Competition held in November 2016.

The problems from the contest are available on p.21.

### Problem 1 Solution

There are 4 small angles

$$\angle A_0 OA_1, \angle A_1 OA_2, \angle A_2 OA_3, \angle A_3 OA_4,$$

3 medium angles

$$\angle A_0 OA_2, \angle A_1 OA_3, \angle A_2 OA_4,$$

and 2 large angles

$$\angle A_0 OA_3, \angle A_1 OA_4,$$

and 1 huge angle

$$\angle A_0 OA_4.$$

Therefore, the number of angles in the diagram is

$$4+3+2+1 = 10.$$

**Answer:** 10

### Problem 2 Solution

Since the ratio of llamas to ostriches is $4 : 7$, 4 out of every

$$4+7 = 11$$

are llamas. Thus, $\frac{4}{11}$ of the 44 total of llamas and ostriches are llamas. Therefore, the petting zoo has

$$\frac{4}{11} \times 44 = 16$$

llamas.

**Answer:** 16

### Problem 3 Solution
The list has 7 numbers, so when we add one more it will have 8. For an average of 5, we need the sum of all 8 numbers to be $8 \times 5 = 40$. Hence we need to insert

$$40 - 1 - 4 - 6 - 4 - 3 - 8 - 2 = 12$$

to get an average of 5.

**Answer:** 12

### Problem 4 Solution
Since $24 = 2^3 \times 3$, the fraction is irreducible if the numerator is not divisible by 2 or 3. There are 8 such numerators:

$$\frac{1}{24}, \frac{5}{24}, \frac{7}{24}, \frac{11}{24}, \frac{13}{24}, \frac{17}{24}, \frac{19}{24}, \frac{23}{24}.$$

**Answer:** 8

### Problem 5 Solution
There are 5 choices for appetizer, $7 + 4 = 11$ choices for entrée, and 3 choices for dessert. Since they can be combined in any way, there are a total of $5 \times 11 \times 3 = 165$ different meals.

**Answer:** 165

### Problem 6 Solution
Note the two shaded regions combine to form a semicircle with radius 6. Thus the area is

$$\frac{1}{2}\pi \times 6^2 = 18\pi$$

so $D = 18$.

**Answer:** 18

### Problem 7 Solution

Joe's mom can clean $\frac{1}{45}$ of the kitchen per minute. When Joe and his mom work together, they can clean $\frac{1}{30}$ of the kitchen per minute. Then, we can find the amount of work that Joe can do per minute

$$\frac{1}{30} - \frac{1}{45} = \frac{1}{90}.$$

Since Joe can clean $\frac{1}{90}$ of the kitchen per minute, it would take Joe 90 minutes to clean the kitchen alone.

**Answer:** 90

### Problem 8 Solution

If all 24 motorcycles are the first kind with a capacity of 2 people, the motorcycles will fit

$$24 \times 2 = 48$$

people, meaning that

$$68 - 48 = 20$$

people will be left out. The second kind of motorcycle can hold

$$3 - 2 = 1$$

extra person, so if we switch 20 of the motorcycles to the second kind everyone will fit. Hence we have

$$24 - 20 = 4$$

motorcycles of the first kind and 20 of the second. Since the first kind costs $40 per motorcycle and the second kind costs $30, the total cost is

$$4 \times 40 + 20 \times 30 = 760$$

dollars to rent the 24 motorcycles.

**Answer:** 760

## Problem 9 Solution

We know $\triangle AFJ$ is isosceles, so

$$\angle AFJ = \angle AJF = (180° - 40°) \div 2 = 70°.$$

Hence using adjacent angles

$$\angle FJI = 180° - 70° = 110°.$$

Noting that the angles in $\triangle BJD$ must add up to $180°$, we have

$$\angle B + \angle D = 180° - 110° = 70°.$$

**Answer:** 70

## Problem 10 Solution

If the lower of the rolls is 1, the other roll could be $1 - 6$ or 6 possibilities. If the lower of the rolls is 2, the other roll could be $2 - 6$ or 5 possibilities. Continuing this way, we see there are

$$6 + 5 + 4 + 3 + 2 + 1 = 21$$

possibilities.

Alternatively, this is the same as the number of ways to put two identical balls into six numbered boxes, so using stars and bars we also see there are

$$\binom{2 + 6 - 1}{6 - 1} = \binom{7}{2} = \frac{7 \times 6}{2} = 21$$

outcomes.

**Answer:** 21

## Problem 11 Solution

Note that only perfect squares have an odd number of factors. The perfect squares between 99 and 500 are

$$10^2 = 100, 11^2 = 121, \ldots, 22^2 = 484.$$

Hence there are $22 - 10 + 1 = 13$ numbers between 99 and 500 that have an odd number of factors.

**Answer:** 13

## Problem 12 Solution

First note that the full ball has surface area

$$4\pi \times 4^2 = 64\pi.$$

For the surface area of the half-ball, we have half of the original surface area

$$64\pi \div 2 = 32\pi$$

but we ALSO have an extra circle of radius 4, with area

$$\pi \times 4^2 = 16\pi.$$

Hence the total surface area of the half-ball is

$$32\pi + 16\pi = 48\pi.$$

**Answer:** 48

## Problem 13 Solution

We don't know the length of the trail, or the time she spent uphill or downhill. However, we may assume that she took 2 hours going uphill. Then the trail length was $3 \times 2 = 6$ miles, and the downhill journey took $6 \div 6 = 1$ hour. Therefore the total time was $2 + 1 = 3$ hours, and the average speed of the whole trip was $(6 + 6) \div 3 = 4$ miles per hour.

**Answer:** 4

### Problem 14 Solution

The last digits of the powers of 2 follow a pattern: 2, 4, 8, 6, 2, 4, 8, 6, ..., which repeats every 4 terms. The exponent 2016 is a multiple of 4 (a full cycle), so the units digit of $2^{2016}$ is the last number in the cycle: 6.

**Answer:** 6

### Problem 15 Solution

There are $2^6 = 64$ total outcomes when you flip a coin 6 times. Since we do not have two heads or two tails in a row, we see the only possibilities are

$$HTHTHT \text{ or } THTHTH.$$

Hence the probability is

$$\frac{2}{64} = \frac{1}{32},$$

so $N = 1, M = 32$ and $N + M = 33$.

**Answer:** 33

### Problem 16 Solution

First note that triangles $\triangle ABC$ and $\triangle ACD$ have the same area so they each have area $400 \div 2 = 200$. Similarly $\triangle ACE$ has the same area as $\triangle CED$, so they each have area $200 \div 2 = 100$.

**Answer:** 100

## Problem 17 Solution

The passenger gets on the bus at 6:16 AM to travel to station B. Since the one-way trip takes 45 minutes, none of the buses from station B have arrived at station A when the passenger departs. The passenger will therefore see every bus that departs station B between 6 AM and their arrival time at station B which is 7:01 AM. One bus leaves exactly at 6 AM, and in the 61 minutes that follow 7 more buses leave, because

$$61 \div 8 \approx 7.625$$

and we round down because the buses leave at the end of each 8 minute interval. Therefore, the passenger sees a total of

$$7 + 1 = 8$$

buses en route from station A to station B.

**Answer:** 8

## Problem 18 Solution

The whole number line has length $14 - 2 = 12$. You win at least as much as you paid if the dial is between 5 and 14 which has length $14 - 5 = 9$. Thus the probability is

$$\frac{9}{12} = \frac{3}{4} = 75\%.$$

**Answer:** 75

## Problem 19 Solution

The only one digit primes are $2, 3, 5, 7$ so the tens digit must be either $2, 3, 5, 7$. Checking we see the two-digit prime-primes are $23, 29, 31, 37, 53, 59, 71, 73, 79$.

**Answer:** 9

## Problem 20 Solution

In the first scenario, each thief gets 8 dollars, but 5 of them receive nothing, so it means they are $8 \times 5 = 40$ dollars short.

In the second scenario, each thief receives 7 dollars, and 5 dollars are left over. Let's give those 5 dollars to 5 of the thieves, one dollar each. We know they are 40 dollars short of giving each thief 8 dollars, so if we add an extra 40 dollars, each thief gets 8 dollars. Since we already gave 5 thieves an extra dollar this means there are $5 + 40 = 45$ thieves in total.

**Answer:** 45

## 2.3  ZIML December 2016 Division M

Below are the solutions from the Division M ZIML Competition held in December 2016.
The problems from the contest are available on p.27.

### Problem 1 Solution
Henry reads 160 pages per day, so in 5 days he reads

$$5 \times 160 = 800$$

pages. Since Henry has $\frac{3}{5}$ of the book remaining the 800 pages he has read are $\frac{2}{5}$ of the book. Therefore, we can divide to get that the book has

$$800 \div \frac{2}{5} = 2000$$

pages in total.

**Answer:** 2000

### Problem 2 Solution
Each smaller rectangle has area 12 cm$^2$. If we compare the top row of 3 horizontal rectangles and the middle row of 4 vertical ones we see the ratio between the length and width of the smaller rectangle is 4 : 3.

Notice that $4 \times 3 = 12$ so in fact each small rectangle has dimensions $4 \times 3$. Thus the base of the big rectangle is

$$4 + 4 + 4 = 12$$

and the height is

$$3 + 4 + 3 = 10$$

so the perimeter is $2 \times (12 + 10) = 44$ cm.

**Answer:** 44

## Problem 3 Solution

Let's say the original price of the item is $100. With the first 30% off, the new price is

$$(1 - .3) \times 100 = 0.70 \times 100 = 70$$

dollars. With another 20% off, the price becomes

$$(1 - .2) \times 70 = 0.80 \times 70 = 56$$

dollars. This is a total discount of

$$100 - 55 = 44$$

dollars, so dividing by the original price the percentage discount is

$$44 \div 100 = 0.44 = 44\%.$$

So the merchant's claim is not correct and the true discount is 44% off the original price.

**Answer:** 44

## Problem 4 Solution

Pump A fills $\frac{1}{12}$ of the pool in one hour, while pump B fills $\frac{1}{18}$ of the pool in one hour. Therefore, the pumps working together fill

$$\frac{1}{12} + \frac{1}{18} = \frac{5}{36}$$

of the pool in one hour. Hence, the two pumps could fill

$$\frac{5}{36} \times 10 = \frac{50}{36}$$

of the pool in 10 hours. Since this overfills the pool by $\frac{7}{18}$, we can turn off one of the pumps for some time. If we want to have both pumps on together for the smallest possible amount of time, we turn off the slower pump (pump B). It takes pump B

$$\frac{7}{18} \div \frac{1}{18} = 7$$

hours to account for the overfilled water, so we can turn pump B off for up to 7 hours. Therefore, the two pumps must work together for at least $10 - 7 = 3$ hours.

**Answer:** 3

## Problem 5 Solution
The last digit must be 0 or 5. Both 1000 and 1005 are not divisible by 11. Considering numbers of the form 10__ __, we see that the smallest possibility is 1045.

**Answer:** 1045

## Problem 6 Solution
Given $\angle DFE = 75°$, we observe that by vertical angles, $\angle AFC = 75°$. Given $\angle BCF = 95°$, we get

$$\angle FCA = 180° - 95° = 85°$$

using supplementary angles. Therefore,

$$\angle CAF = 180° - 75° - 85° = 20°.$$

**Answer:** 20

## Problem 7 Solution
There are 80 more bird legs than cat legs. Since each bird has 2 legs, if we remove
$$80 \div 2 = 40$$

birds then we are left with

$$100 - 40 = 60$$

animals and there are an equal number of bird and cat legs. Since each cat has twice as many legs as a bird, there must be half as many birds as cats. Therefore there are

$$60 \div 3 = 20$$

cats and

$$20 \times 2 = 40$$

of the remaining animals are birds. Hence there are

$$40 + 40 = 80$$

birds in total.

**Answer:** 80

## Problem 8 Solution
The shaded region is a semicircle of radius 3, plus a semicircle of radius 4, plus a right triangle with legs 6 and 8, minus a semicircle of radius 5. Hence the area is

$$\frac{1}{2}\pi \times 3^2 + \frac{1}{2}\pi \times 4^2 + \frac{1}{2} \times 6 \times 8 - \frac{1}{2}\pi \times 5^2 = 24.$$

**Answer:** 24

## Problem 9 Solution
Note that there are two ways of going from $A$ to $A$. Namely, one can take the $A \rightarrow B \rightarrow A$ path or the $A \rightarrow B \rightarrow C \rightarrow A$ path.

For the first case, there are 5 choices of paths to take from $A$ to $B$. Per choice, there are 4 choices of paths to take from $B$ to $A$. Therefore, there are

$$5 \times 4 = 20$$

different paths to take without going to $C$. The second case allows one to go to $C$ which can be achieved in

$$5 \times 3 = 15$$

ways. Since there are 2 paths from $C$ to $A$, there are

$$15 \times 2 = 30$$

total ways to get from $A$ to $A$ going through $C$. Therefore, there are

$$20 + 30 = 50$$

possible paths from $A$ to $A$ without visiting any place twice and all roads are one-way.

**Answer:** 50

## Problem 10 Solution

Note that $30 = 5 + 10 + 10 + 5$. Mr. Tahoe could have seen the last 5 minutes of one presentation, then 2 whole presentations, and then the first 5 minutes of the next presentation. So, Mr. Tahoe could have seen 4 presentations.

**Answer:** 4

## Problem 11 Solution

Note that $4000 = 2^5 \times 5^3$. Therefore the perfect squares are $2^0 \times 5^0 = 1$, $2^2 \times 5^0 = 4$, $2^4 \times 5^0 = 16$, $2^0 \times 5^2 = 25$, $2^2 \times 5^2 = 100$, and $2^4 \times 5^2 = 400$. Thus there are 6 total factors that are also perfect squares.

**Answer:** 6

## Problem 12 Solution

We note that the bag contains objects defined by its shape (cubes vs. balls) and color (red vs. blue). Shape and color are independent from each other, so one can interpret the complement of

cubed-shaped objects to be ball-shaped objects and the complement of red objects to be blue.

Since we are interested in minimizing the number of red balls in
the bag, we want to maximize the number of red cubes in the bag.
Given that there is at least one blue cube in the bag, the maximum
number of red cubes is

$$20 - 1 = 19.$$

Since there are 40 total red objects, the minimum number of red
balls is
$$40 - 19 = 21.$$

**Answer:** 21

### Problem 13 Solution
It is the same as the remainder when we add up $1 + 2 + 3 + \cdots +$
2016, whose result is $\dfrac{2016 \times 2017}{2} = 1008 \times 2017$, which is a
multiple of 9. So the answer is 0.

**Answer:** 0

### Problem 14 Solution
Consider the shortest path (a straight line) on the unfolded cube
shown below, where $C$ is the vertex directly below $B$:

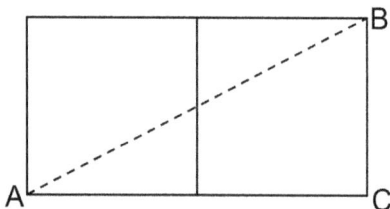

In this diagram, $ABC$ is a right triangle, so

$$AB^2 = AC^2 + BC^2 = 4^2 + 2^2 = 20.$$

Therefore the shortest distance from $A$ to $B$ along the surface of the cube is $\sqrt{20}$ so $K = 20$.

**Answer:** 20

## Problem 15 Solution

There are 40 cards in total. Hence there are $40 \times 39$ ways to be dealt a first card and a second card. The first card can be any of the 40. However, once the first card is chosen, there are $10 - 1 = 9$ remaining cards that have the same color. Thus there are $40 \times 9$ ways where both cards are the same color. The probability is then

$$\frac{40 \times 9}{40 \times 39} = \frac{9}{39} = \frac{3}{13},$$

so $N + M = 3 + 13 = 16$.

**Answer:** 16

## Problem 16 Solution

Call the entire triangle $ADE$ (with $C$ on $\overline{AE}$). $\triangle ABC$ and $\triangle ABE$ have the same height, so

$$[ABC] : [ABE] = 3 : 5 = 12 : 20.$$

Similarly,

$$[ABE] : [ADE] = 4 : 5 = 20 : 25.$$

Combining these two we have

$$[ABC] : [ADE] = 12 : 25.$$

As $[ADE] = 25$, we have $[ABC] = 12$.

**Answer:** 12

## Problem 17 Solution

We know it takes 3 hours and 50 minutes, or

$$3 \times 60 + 50 = 230$$

minutes to travel uphill. We want to find how much of this time Sam was actually walking. A walking/resting cycles takes 40 minutes, so because

$$230 = 5 \times 40 + 30$$

the 230 minutes will consist of 5 walking/resting cycles and end with Sam walking the final 30 minutes. Hence, he walks a total of

$$5 \times 30 + 30 = 180$$

minutes. Since Sam walks downhill at a speed 1.5 times faster as that he walks uphill, he take will spend 1.5 times less time walking downhill, a total of

$$180 \div 1.5 = 120$$

minutes. In these 120 minutes, me must rest

$$(120 \div 30) - 1 = 3$$

times (since he can walk the final 30 minutes). He therefore spends a total of

$$120 + 3 \times 5 = 135$$

minutes traveling downhill.

**Answer:** 135

## Problem 18 Solution

Note that if you know the sum of the 5 numbers, you know the average. Therefore we need to count the possibilities for the sum of the 5 numbers. The smallest sum is

$$1 + 1 + 1 + 1 + 1 = 5$$

and the largest sum is

$$10 + 10 + 10 + 10 + 10 = 50.$$

Note that all of the sums in between $(6, 7, 8, \ldots, 50)$ are all possible. Hence there are

$$50 - 5 + 1 = 46$$

possible sums and therefore 46 possible averages.

**Answer:** 46

### Problem 19 Solution
$90 = 2 \times 3^2 \times 5$. Also, $180 = 2^2 \times 3^2 \times 5$ and $30 = 2 \times 3 \times 5$. If we denote the other number as $2^x \times 3^y \times 5^z$, we get (by comparing the exponents of each prime number)

$$x = 2, y = 1, z = 1.$$

So the other number is $2^2 \times 3 \times 5 = 60$.

**Answer:** 60

### Problem 20 Solution
Notice that the sequence is

$$0^3 + 1, 1^3 + 1, 2^3 + 1, 3^3 + 1, 4^3 + 1, 5^3 + 1,$$

so the next term is
$$6^3 + 1 = 217.$$

**Answer:** 217

## 2.4  ZIML January 2017 Division M

Below are the solutions from the Division M ZIML Competition held in January 2017.
The problems from the contest are available on p.33.

### Problem 1 Solution
We can use trial and error:
$$\frac{5}{11} = \frac{10}{22} = \frac{15}{33} = \frac{20}{44} = \frac{25}{55} = \frac{30}{66} = \cdots$$
and we see that $m = 25, n = 55$ works. Note here that the sum of the numerator and denominator increases by $5 + 11 = 16$ each 'step'.

**Answer:** 25

### Problem 2 Solution
Extend the side of the smaller rectangle to cut the bigger rectangle and form a tiny square as shown.

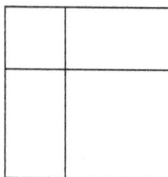

The tiny square has area
$$44 - 28 = 16$$
and hence a side length of 4. The small rectangle thus has dimensions $4 \times 7$ and the smaller square is 7 by 7 with area $7^2 = 49$.

**Answer:** 49

## Problem 3 Solution

To be divisible by 5, the number must end with a 5, and to be divisible by 3, the sum of the digits must be divisible by 3. This means that the number of 5's is a multiple of 3. The smallest multiple of 3 is 3, so the answer has one 3 and three 5's. We want the smallest such number, which is 3555.

**Answer:** 3555

## Problem 4 Solution

The number 7 is a prime number, so the only factors of 7 are 1 and 7. Therefore, when Paul runs the program the first time, the output is

$$1 + 7 = 8.$$

8 has factors

$$1, 2, 4, 8$$

so when he runs it a second time the output is

$$1 + 2 + 4 + 8 = 15.$$

Hence the answer is 15.

**Answer:** 15

## Problem 5 Solution

A full batch can be carried by 36 big trucks. Since each big truck can load 4 more tons than each small truck, if 36 small trucks are used, there are still

$$36 \times 4 = 144$$

tons of steel blocks left over. However, we know that 45 small trucks in total can carry the full batch, so the

$$45 - 36 = 9$$

extra trucks must carry the extra 144 tons. Hence each small truck can carry

$$144 \div 9 = 16$$

tons of steel blocks. Finally, this means that one batch, which is carried by 45 small trucks, is a total of

$$45 \times 16 = 720$$

tons of steel blocks.

**Answer:** 720

## Problem 6 Solution

Suppose that the two people order burgers. Then the first person has a choice between 7 burgers. Because the two people are required to order different things, the second person has a choice between 6 burgers. Therefore, there are

$$7 \times 6 = 42$$

different ways for two people to order burgers. Similarly, for salads, the first person has a choice between 4 salads and the second person has a choice between 3 salads. Therefore, there are

$$4 \times 3 = 12$$

different ways for two people to order salads. Therefore, there is a total of

$$42 + 12 = 54$$

different ways for two people to order burgers or salads without ordering the same kind of food.

**Answer:** 54

## Problem 7 Solution

Bella can buy 3 more shawls than belts. These 3 shawls cost a total of

$$3 \times 12 = 36$$

dollars in total. Every shawl is

$$14 - 12 = 2$$

dollars cheaper than a belt. Hence if Bella buys

$$36 \div 2 = 18$$

shawls instead of 18 belts, she will have enough leftover money to buy 3 extra shawls. Hence Bella has exactly enough money to buy 18 belts, which is

$$18 \times 14 = 252$$

in total.

**Answer:** 252

## Problem 8 Solution

The area of the full circle is

$$4^2 \pi = 16\pi.$$

Therefore, the shaded region is

$$\frac{14}{16} = \frac{7}{8}$$

of the full circle, so the angle at point $O$ is

$$\frac{360°}{8} = 45°.$$

So the triangle is an isosceles right triangle, and its area is

$$4^2 \div 2 = 8.$$

**Answer:** 8

## Problem 9 Solution

Andy takes $\frac{1}{3}$ more time to plant a tree. Since no time is mentioned at all in the problem, we may assume it takes Nathan 3 minutes to plant a tree, so it takes 4 minutes for Andy to plant a tree. Therefore, in 12 minutes, Nathan can plant 1 more tree than Andy (as Andy can plant 3 trees and Nathan can plant 4 trees). Therefore, in

$$12 \times 36 = 432$$

minutes, Nathan plants 36 more trees than Andy. In this time Nathan plants

$$4 \times 36 = 144$$

trees and Andy plants

$$3 \times 36 = 108$$

trees. A total of $144 + 108 = 252$ trees are planted.

**Answer:** 252

## Problem 10 Solution
Since the balls are picked at once, the total number of outcomes in the sample space is

$$\binom{13}{5} = 1287$$

since we are interested in choosing 5 balls from a group of 13 balls. We have two cases to consider.

The first case is when all 5 balls are red. There are

$$\binom{5}{5} = 1$$

ways of choosing 5 balls to be red.

The second case is when all 5 balls are green. There are

$$\binom{8}{5} = 56$$

ways of choosing 5 balls to be green. Therefore, the desired probability is

$$\frac{\binom{5}{5}+\binom{8}{5}}{\binom{13}{5}}=\frac{57}{1287}=\frac{19}{429}.$$

Thus $Q = 429$ and $P = 19$ so $Q - P = 410$.

**Answer:** 410

**Problem 11 Solution**

First note since we are attaching squares to a regular pentagon, all the marked segments in the diagram below are equal.

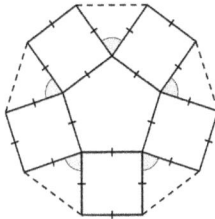

Therefore all the outside triangles are isosceles. Further, all the marked angles are equal, with measure

$$360° - 90° - 108° - 90° = 72°.$$

Hence these triangles have one angle of $72°$ and the other two angles are

$$(180° - 72°) \div 2 = 54°.$$

Thus the triangles have angles $72°, 54°, 54°$ so the difference between the largest and smallest angles is $72 - 54 = 18$.

**Answer:** 18

**Problem 12 Solution**

Since they are traveling in the same direction, their relative speed is the difference of the two speeds,

$$12 - 8 = 4 \text{ miles per hour}$$

towards each other before passing, and away from each other after passing. Emerson starts a half mile in front of Emily, and ends a half mile behind Emily, so Emily travels a total of 1 mile in reference to Emerson. Therefore, the time Emerson is in her view is

$$1 \div 4 = \frac{1}{4}$$

of an hour, which is 15 minutes.

**Answer:** 15

**Problem 13 Solution**

First use the Pythagorean theorem to find the distance between opposite corners of the base (i.e. the length of the diagonal of the rectangular base):

$$\sqrt{3^2 + 4^2} = 5.$$

Note then this diagonal is perpendicular to the vertical edges of the rectangular prism, so we can use the Pythagorean theorem again to find the distance between opposite corners:

$$\sqrt{5^2 + 12^2} = 13.$$

Therefore the length of the line connecting the two corners is 13.

**Answer:** 13

**Problem 14 Solution**

Say the digits are $a, b, c, d$ so the numbers are of the form $\overline{abcd}$ with

$$1 \leq a, b, c, d \leq 9 \text{ and } a+b+c+d = 8.$$

Note that because $a+b+c+d=8$ we automatically have

$$a,b,c,d \leq 9.$$

Hence we need

$$a+b+c+d=8 \text{ with } a,b,c,d \geq 1$$

so we can use the positive version of stars and bars. Hence there are

$$\binom{4+4-1}{4} = 35$$

possible numbers.

**Answer:** 35

### Problem 15 Solution

Given that there are 50 students that scored above 600 on either the critical reading or math section of the SAT, the overlap of students that score above 600 on either sections represent the number of students that score above 600 on both sections. Therefore, there are

$$(35+25) - 50 = 10$$

students that score above 600 on both critical reading and math sections of the SAT.

The maximum total score that these 10 students can score is

$$800 + 800 = 1600,$$

and the maximum total score that the remaining 40 students can score is

$$800 + 600 = 1400.$$

Therefore, the sum of the 50 SAT total scores is

$$40 \times 1400 + 10 \times 1600 = 72000$$

and thus,

$$72000 \div 50 = 1440$$

is the average of the 50 SAT scores.

**Answer:** 1440

## Problem 16 Solution

We know there are

$$100 + 200 = 300$$

total students. We need to know how many students are in grade 6. 11% of the 100 students in school A are in grade 6, so

$$11\% \times 100 = 11$$

students in school A are grade 6. Similarly,

$$17\% \times 200 = 34$$

students are in grade 6 from school B. This gives a total of

$$11 + 34 = 45$$

students in grade 6. Therefore,

$$\frac{45}{300} = 0.15 = 15\%$$

of the students would be in grade 6 if the two schools combined.

**Answer:** 15

## Problem 17 Solution

The denominator

$$1155 = 3 \cdot 5 \cdot 7 \cdot 11$$

must be the common denominator of the four consecutive primes, so the primes are 3, 5, 7, and 11. Then calculating we have

$$\frac{1}{3} + \frac{1}{5} + \frac{1}{7} + \frac{1}{11} = \frac{385 + 231 + 165 + 105}{1155} = \frac{886}{1155},$$

so $N = 886$.

**Answer:** 886

## Problem 18 Solution
The big circle has radius 3. The perimeter of the shaded region equals the sum of the perimeters of all the circles. Therefore the answer is

$$7 \times (2\pi) + (2 \times 3)\pi = 20\pi.$$

**Answer:** 20

## Problem 19 Solution
The total length of the number line is $7 - (-3) = 10$. Note the number squared is greater than 4 if the number is greater than 2 or less than $-2$. The length of the number line consisting of all numbers greater than 2 is

$$7 - 2 = 5$$

and the length of the number line consisting of all numbers less than $-2$ is

$$-2 - (-3) = 1.$$

Therefore, the length of numbers such that when it is squared, the value is greater than 4 is

$$5 + 1 = 6$$

so the desired probability is

$$\frac{6}{10} = 60\%$$

so $K = 60$.

**Answer:** 60

## Problem 20 Solution

$d$ is a divisor of $22 + 41 + 60 - 21 = 102 = 2 \times 3 \times 17$. If we want the three remainders $r_1, r_2, r_3$ to add up to 21, $d$ cannot be a small number like 2, or 3, or 6. But if $d$ is too big, such as $2 \times 17 = 34$, then the number 22 is already its own remainder, and greater than 21. The only possibility is $d = 17$. Verify: $22 \equiv 5 \pmod{17}$, $41 \equiv 7 \pmod{17}$, and $60 \equiv 9 \pmod{17}$. The three remainders $5 + 7 + 9 = 21$ exactly.

**Answer:** 17

## 2.5   ZIML February 2017 Division M

Below are the solutions from the Division M ZIML Competition held in February 2017.
The problems from the contest are available on p.39.

### Problem 1 Solution
First find the current pure gold amount, which is

$$10 \times 40\% = 4$$

grams. Therefore the non-gold portion of the alloy is 6 grams. This non-gold quantity does not change after adding more pure gold, and it will be $1 - 60\% = 40\%$ of the new alloy. So the total weight of the new alloy is $6 \div 40\% = 15$ grams. That means $15 - 10 = 5$ grams of pure gold should be added.

**Answer:** 5

### Problem 2 Solution
We first calculate the speed of the original train. Note to completely pass through the tunnel the train must travel the length of the tunnel plus the length of the train. Since the length of the train does not change and it takes 25 seconds to pass through a 250 meter tunnel and 23 seconds to pass through a 210 meter tunnel it must travel the extra

$$250 - 210 = 40$$

meters in

$$25 - 23 = 2$$

seconds. Therefore the speed of the train is

$$40 \div 2 = 20$$

m/s. Therefore, in the 25 seconds it takes to pass through the 250 meter tunnel, the train travels

$$20 \times 25 = 500$$

meters. Since the tunnel itself is 250 meters long, the difference

$$500 - 250 = 250$$

meters must be the length of the original train.

We can now find how long it takes to pass the approaching train. To pass each other, the trains must travel a distance of

$$320 + 250 = 570$$

meters (the sum of the lengths of both trains). Since the trains are approaching each other, their relative speed is

$$18 + 20 = 38$$

m/s. Therefore, it takes

$$570 \div 38 = 15$$

seconds for the original train to pass the approaching train.

**Answer:** 15

**Problem 3 Solution**

If we first assume all 70 loads are all from Truck B, there would be a total of

$$70 \times 4 = 280$$

cubic meters of dirt removed. In actuality,

$$400 - 280 = 120$$

more cubic meters of dirt were removed. This means that Truck A must have taken some of the loads of dirt. Each load from Truck A has

$$7 - 4 = 3$$

more cubic meters than that from Truck B, so if we take

$$120 \div 3 = 40$$

loads from truck B and give them to Truck A, we will have the correct amount of dirt. Hence truck A carries 40 loads of dirt.

**Answer:** 40

## Problem 4 Solution

The ratio of packs of crayons to students is $2 : 5$. Since there are 18 packs of crayons available, if we multiply both sides of the ratio by

$$18 \div 2 = 9$$

we have

$$2 : 5 = 18 : 45.$$

So, 45 students can share 18 packs of crayons.

**Answer:** 45

## Problem 5 Solution

Alice travels 60 mph, so she passes the car after she travels

$$60 \times 6 = 360$$

miles. Similarly, Bob passes the car after traveling

$$48 \times 7 = 336$$

miles. Therefore, the car moving the opposite direction traveled

$$360 - 336 = 24$$

miles in one hour, hence is traveling 24 mph. Therefore, after another hour, the car traveling the opposite direction will be

$$336 - 24 = 312$$

miles from site A when Cindy passes it. Since Cindy has been traveling 8 hours when this happens, Cindy's speed is

$$312 \div 8 = 39$$

mph.

**Answer:** 39

## Problem 6 Solution

The hexagon is made up of 6 equilateral triangles, each with an area of $150\sqrt{3} \div 6 = 25\sqrt{3}$. Let $s$ be the side length of an equilateral triangle, the height of the triangle is then $\frac{s\sqrt{3}}{2}$, and it has area

$$\frac{s^2\sqrt{3}}{4} = 25\sqrt{3}.$$

Thus, the side length of the hexagon is $s = 10$.

**Answer:** 10

## Problem 7 Solution

$\triangle BCP$ is isosceles, with $\angle BCP = 90° + 60° = 150°$, so

$$\angle CBP = \angle BPC = (180° - 150°) \div 2 = 15°.$$

Similarly, $\angle BAQ = \angle BQA = 15°$. Hence

$$\angle ABP = 90° - 15° = 75°,$$

and $\angle AGP = \angle AGB = 180° - 15° - 75° = 90°$.

**Answer:** 90

## Problem 8 Solution

Using the Pythagorean theorem $AC^2 = AB^2 + BC^2$, so $AC = 10$ (recall $(6, 8, 10)$ is a Pythagorean triple). Using the Pythagorean theorem again $AD^2 = AB^2 + BD^2$, so $AD = 17$ (as $(8, 15, 17)$ is another Pythagorean triple). The perimeter of $\triangle ACD$ is $AC + AD + CD = 10 + 17 + 9 = 36$.

**Answer:** 36

## Problem 9 Solution

The areas of $\triangle AOB$ and $\triangle DOC$ add up to half of the area of the rectangle $ABCD$. Therefore, the rectangle area is $2(6 + 12) = 36$ cm$^2$.

**Answer:** 36

## Problem 10 Solution

Since $\triangle BCE$ and $\triangle AFE$ are $45 - 45 - 90$ triangles and $A$ is the midpoint of $\overline{BE}$, we get that $F$ is the midpoint of $\overline{AD}$. From here we get $[ABCF] = \frac{3}{4} \cdot [ABCD]$, and $[DFC] = \frac{1}{4} \cdot [ABCD]$, so $[ABCF] - [DFC] = \frac{1}{2} \cdot [ABCD]$. Hence $[ABCD] = 8$.

**Answer:** 8

## Problem 11 Solution

Alice and Bob have 3 options for couches. Per couch option, there are 6 coffee tables to choose from. Per couch to coffee table pairing, there are 5 recliners to choose from. Therefore, there are

$$6 \times 3 \times 5 = 90$$

possible different furniture combinations that Alice and Bob can make.

**Answer:** 90

## Problem 12 Solution

This is equivalent to placing 5 identical balls into 3 different boxes (one box for each type of chocolate, and as many chocolates of a kind as balls in its box). Thus, using stars and bars we have there are $\binom{5+3-1}{5} = 21$ different ways to choose which chocolates to get.

**Answer:** 21

## Problem 13 Solution

Assuming that the letters are all distinct, there are

$$7! = 7 \times 6 \times 5 \times 4 \times 3 \times 2 \times 1 = 5040$$

ways to arrange 7 letters to form words. However, since there are three $E$'s, we need to remove all of the duplicate arrangements by dividing by the number of ways to arrange the three $E$'s. This can be done in

$$3! = 3 \times 2 \times 1 = 6$$

ways. Therefore, there are

$$5040 \div 6 = 840$$

ways to rearrange the letters in ARETEEM.

**Answer:** 840

## Problem 14 Solution

In the faculty of the mathematics department, there are

$$\binom{14}{5} = 2002$$

ways to form a committee of 5 people. We will use complementary counting to count the number of possible committees containing at least 2 women. Note that there are

$$\binom{6}{1} = 6$$

ways to choose 1 women and

$$\binom{8}{4} = 70$$

ways to choose 4. Therefore, for the case of a committee with exactly 1 woman, there are

$$70 \times 6 = 420$$

ways that this can be done. For the case with no women on the committee, this is done in

$$\binom{8}{5} = 56$$

ways. Therefore, there are

$$2002 - 420 - 56 = 1526$$

ways of forming a 5−member committee with at least 2 women.

**Answer:** 1526

## Problem 15 Solution

As we have at most 5 of a single bill, counting the possible monetary values is equivalent to find all non-negative solutions to the equation $a+b+c+d+e = 5$. Using stars and bars we get there are $\binom{5+4-1}{5} = 56$ different possible monetary values in the box.

**Answer:** 56

## Problem 16 Solution

The number of students that passed must be a common multiple of 3 and 4, that is, a multiple of $\mathrm{lcm}(3,4) = 12$. The only multiple of 12 less than or equal to 20 is 12 itself, so that must be the number of students that passed the test.

**Answer:** 12

## Problem 17 Solution
The digit a has to be even. The sum of digits $a + 2 + a = 2 + 2a$ is divisible by 3. The only two possibilities are $a = 2$ and $a = 8$. Therefore, the greatest possible 5−digit number is 35828.

**Answer:** 35828

## Problem 18 Solution
Only numbers that end in 1, 5 or 6 end with the same digit when squared. A number that ends in 1 is of the form $10a + 1$, and

$$(10a + 1)^2 = 120a + 1.$$

Note the tens digit of this square number will be the same as the ones digit of $2a$, so the only number that would work is $a = 0$, which does not yield a 2-digit number. A number that ends in 5 is of the form $10a + 5$, and

$$(10a + 5)^2 = 200a + 25,$$

that is, it always ends in 25; so 25 is the only 2-digit number ending in 5 that ends with the same two digits when squared. A number that ends in 6 is of the form $10a + 6$, and

$$(10a + 6)^2 = 220a + 36.$$

Note the tens digit of this square number will be 3 more than the ones digit of $2a$. Noticing this pattern, we see the only number that works is $a = 7$, so 76 is the only other 2-digit number that ends with the same two digits when squared.

**Answer:** 76

**Problem 19 Solution**

The units digit of the first few powers of 7 are 7, 9, 3, 1, 7, 9, 3, 1,..., which follows a cycle of length 4. Hence the units digit of $7^{47}$ is the same as $7^3$, which is 3, and the units digit of $7^{74}$ is the same as $7^2$, which is 9. Hence the sum has units digit 2.

**Answer:** 2

**Problem 20 Solution**

Note that we can rewrite the expression as $13^{234}(1 + 13 + 13^2)$. Since $1 + 13 + 13^2 = 183 = 3 \cdot 61$ we have the prime factorization of $13^{234} + 13^{235} + 13^{236} = 3 \cdot 13^{234} \cdot 61$ so the greatest prime factor is 61.

**Answer:** 61

## 2.6 ZIML March 2017 Division M

Below are the solutions from the Division M ZIML Competition held in March 2017.

The problems from the contest are available on p.45.

### Problem 1 Solution

The area of the shaded region is

$$40 - 5 - 20 = 15.$$

**Answer:** 15

### Problem 2 Solution

Since each team gets one home game and one away game versus a second team, the order of the arrangements of the teams matters. Therefore, we have a permutation. Namely, the number of possible games in the tournament is

$$_8P_2 = \frac{8!}{6!} = 8 \times 7 = 56.$$

**Answer:** 56

### Problem 3 Solution

The ratio of red to white marbles they both have is $20 : 16 = 5 : 4$. So, for every $4 + 5 = 9$ marbles Rosa has, 4 are white. That is, Rosa has $27 \times \dfrac{4}{9} = 12$ white marbles.

**Answer:** 12

### Problem 4 Solution

The area $A$ minus the area $B$ is equal to the semicircle minus the triangle:

$$\frac{10^2 \pi}{2} - \frac{20 \times 10}{2} = 50\pi - 100.$$

Using $\pi \approx 3.14$ we have

$$50\pi - 100 = 157 - 100 = 57.$$

**Answer:** 57

## Problem 5 Solution

The only requirement for participants to advance to the next race is if they are placed in the top 4. Therefore, there are

$$\binom{8}{4} = 70$$

ways of choosing 4 participants out of the 8 to finish in the top 4.

**Answer:** 70

## Problem 6 Solution

If we subtract 2 from this number, the resulting number is a common multiple of 4 and 6. Since $\text{lcm}(4,6) = 12$, we are looking for the largest multiple of 12 that has two digits after adding 2. Now, $96 = 8 \times 12$ is the largest two digit multiple of 12, and adding 2 to this number gives 98.

**Answer:** 98

## Problem 7 Solution

Since Stephanie and Bob are traveling in opposite directions, their relative speed is the sum of the two speeds, which is, $4 + 6 = 10$ km per hour towards each other. Thus, it takes them $50 \div 10 = 5$ hours to pass each other on the trail.

**Answer:** 5

## Problem 8 Solution

The prime factorizations of 2016 and 2160 are $2016 = 2^5 \times 3^2 \times 7$, and $2160 = 2^4 \times 3^3 \times 5$, so

$$\gcd(2016, 2160) = 2^4 \times 3^2 = 144.$$

Hence, we are looking for 2-digit factors of 144. These factors are 12, 16, 18, 24, 36, 48, and 72. There are 7 factors.

**Answer:** 7

### Problem 9 Solution

We know $\triangle AFJ$ is isosceles, so

$$\angle AFJ = \angle AJF = (180° - 30°) \div 2 = 75°.$$

Hence using adjacent angles

$$\angle FJI = 180° - 75° = 105°.$$

Noting that the angles in $\triangle BJD$ must add up to $180°$, we have

$$\angle B + \angle D = 180° - 105° = 75°.$$

**Answer:** 75

### Problem 10 Solution

Since the only prime number between 20 and 25 is 23 we need to check which of the numbers $230, 231, 232, \ldots, 239$ are prime. Clearly the even numbers are not, 231 and 237 are divisible by 3, and 235 is divisible by 5. In fact 233 and 239 are prime, so the answer is 2.

**Answer:** 2

### Problem 11 Solution

Note since the number is divisible by 3 and 5, it will be divisible by 15. We are looking for numbers between 201700 and 201799. There are $\left\lfloor \dfrac{201699}{15} \right\rfloor = 13446$ multiples of 15 less than 201699 and $\lfloor\dfrac{201799}{15}\rfloor = 13453$ multiples of 15 less than 201800, so there are $13453 - 13446 = 7$ numbers.

**Answer:** 7

### Problem 12 Solution

For the red checker piece, there are 64 possible options. This checker piece can be placed anywhere on the checkerboard. After this checker piece is placed, the black checker piece must lie on the square with a color different than the color of the square where the first checker piece lies on. There are 32 options for the second checker piece.

Therefore, there are a total of

$$64 \times 32 = 2048$$

ways to do this.

**Answer:** 2048

### Problem 13 Solution

Since the distance is the same in both directions, we can assume that the distance is a value that is easy to deal with. Therefore, assume the distance between Irvine and Fullerton is 120 miles. So the time takes me to reach Fullerton is $120 \div 60 = 2$ hours, and the return trip takes $120 \div 40 = 3$ hours. For the round trip, the total distance is $120 + 120 = 240$ miles and the total time spent is $2 + 3 = 5$ hours. Thus, the average speed is $240 \div 5 = 48$ miles per hour.

**Answer:** 48

### Problem 14 Solution

If all 16 packs were colored pencils, then the teacher would have spent $19 \times 16 = 304$ dollars, which is $304 - 280 = 24$ more dollars than the actual amount. Each pack of regular pencils is $19 - 11 = 8$ dollars less expensive than colored pencils. Therefore the math teacher bought $24 \div 8 = 3$ packs of regular pencils and $16 - 3 = 13$ packs of colored pencils.

**Answer:** 13

### Problem 15 Solution

First use the Pythagorean theorem to find the distance between opposite corners of one of the faces:

$$\sqrt{9^2 + 12^2} = 15.$$

Note then that this diagonal is perpendicular to the vertical edges of the rectangular prism, so we can use the Pythagorean theorem again to find the distance between opposite corners:

$$\sqrt{8^2 + 15^2} = 17.$$

Therefore the length of the line connecting the two corners is 17.

**Answer:** 17

### Problem 16 Solution

If the length of one of the sticks is shorter than $\frac{1}{2}$, then the other stick is greater than $\frac{3}{2}$ and with these lengths, it is impossible to form a triangle. Hence, as the stick has length 2, we must break it at least .5 units from each end. This gives us a probability of $\frac{1}{2}$. Hence $p + q = 1 + 2 = 3$.

**Answer:** 3

### Problem 17 Solution

The first ammonia solution has $15 - 10 = 5$ percent less ammonia than the target concentration. the second ammonia solution has

$30 - 15 = 15$ more ammonia than the target concentration. The ratio of this differences is $5 : 15 = 1 : 3$. Thus, in order to reach the target concentration we need the number of gallons of each solution to be in the opposite ratio. As $3 : 1 = 150 : 50$, we need to mix 150 gallons of the 10% ammonia solution.

**Answer:** 150

**Problem 18 Solution**
Let $O$ be the feet of the altitude from $A$.

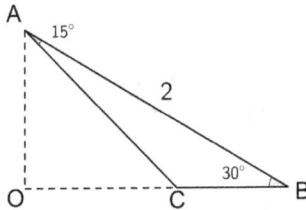

Since $\angle ACB = 180° - 15° - 30° = 45°$, $\triangle AOC$ is a $45 - 45 - 90$ triangle and $\triangle AOC$ is a $30 - 60 - 90$ triangle. The sides of a $30 - 60 - 90$ triangle are in ratio $1 : \sqrt{3} : 2$, so $AO = 1$ and $BO = \sqrt{3}$. The sides of a $45 - 45 - 90$ triangle are in ratio $1 : 1 : \sqrt{2}$, so $OC = 1$. Thus, the area of $\triangle ABC$ is

$$\frac{\sqrt{3}}{2} - \frac{1}{2} = \frac{-1 + \sqrt{3}}{2},$$

thus $P + Q + R = 4$.

**Answer:** 4

**Problem 19 Solution**
When we multiply a number by 11, the last digit of the product is the same as the last digit of the original number, and the tens digit will be the sum of the ones and tens digits of the original number. This means that the original number has 1 one and

$4 - 1 = 3$ tens. Since the product starts with 3, that means our original number starts with either 2 or 3. The digit we are missing will be the sum of the tens and thousands digit of our original number. If the original number starts with 2, it would be 231, but $231 \times 11 = 2541$. If the original number starts with 3, it would be 331 and $331 \times 11 = 3641$, so the missing digit on the wall is 6.

**Answer:** 6

## Problem 20 Solution

Note the question is equivalent to putting 3 identical balls into 8 labeled boxes, where the first box has at least one ball. This means we need to put 2 balls into 8 boxes (without restrictions). For this we can use stars and bars, with 2 stars and $8 - 1 = 7$ bars. Hence there are

$$\binom{2+8-1}{2} = \binom{9}{2} = \binom{9}{7} = 36$$

outcomes.

**Answer:** 36

## 2.7   ZIML April 2017 Division M

Below are the solutions from the Division M ZIML Competition held in April 2017.
The problems from the contest are available on p.51.

### Problem 1 Solution
There are currently 7 numbers that add up to 28. When we add a new number, to get an average of 5, we need the 8 numbers to sum to $8 \times 5 = 40$. Therefore, the new number must be $40 - 28 = 12$.

**Answer:** 12

### Problem 2 Solution
Consider the $5 \times 5$ square that encloses quadrilateral $ABCD$. Note that $\overline{AC}$, $\overline{BD}$ divide the entire square into four rectangles. For each of those rectangles, half of the rectangle is shaded, hence the area of $ABCD$ is half of the area of the entire grid, that is,

$$[ABCD] = \frac{5^2}{2} = \frac{25}{2}.$$

**Answer:** 12.5

### Problem 3 Solution
Jim earns 7% commission on the first $700 of sales, a total of

$$7\% \times 700 = 0.07 \times 700 = 49$$

dollars. He earns 12% on the additional

$$1200 - 700 = 500$$

dollars of sales, for an additional commission of

$$12\% \times 500 = 0.12 \times 500 = 60$$

dollars. In total, Jim's earns

$$49 + 60 = 109$$

dollars as commission.

**Answer:** 109

### Problem 4 Solution

The side length of the bigger square is 3 times the side length of the smaller square, so we see that 9 of the little squares make up the larger square. The area of the bigger square is 24 square inches more than the smaller square, which corresponds to 8 extra small squares, so the area of a single small square is

$$24 \div 8 = 3$$

square inches. As the whole rectangle is made up of 12 small squares, it has an area of $12 \times 3 = 36$ square inches.

**Answer:** 36

### Problem 5 Solution

A single run/rest period takes $6 + 2 = 8$ minutes. Since

$$26 \div 8 = 3 \, R \, 2,$$

we see that Sam rests a total of $3 \times 2 = 6$ minutes, so he runs a total of 20 minutes. (After the third rest, Sam runs the last 2 minutes to finish the 5 km.) Therefore, Sam runs at a rate of 5 km per 20 minutes, or 15 km per 60 minutes. Hence Sam runs at 15 kilometers per hour.

**Answer:** 15

### Problem 6 Solution

We are given that the ratio of the amount of work each can complete in a given time is

$$\frac{1}{2} : \frac{1}{3} = 3 : 2.$$

Hence, working for the 15 days, Emily completed $\frac{3}{5}$ of the project (and Julia completed the other $\frac{2}{5}$ of the project). Hence Emily can complete

$$\frac{3}{5} \div 15 = \frac{1}{25}$$

of the project in one day, so it would take her 25 days to finish the project by herself.

**Answer:** 25

## Problem 7 Solution

Note we only need to worry about the units digits, so we try to find the units digit of $4^{200}$ and $6^{300}$. For powers of 4, the units digits follow a pattern 4, 6, 4, 6, …. Powers of 6 always end in 6. Since 200 is even, $14^{200}$ will have a units digit 6, and the units digit of $16^{300}$ is also 6. Since $6 + 6 = 12$ has units digit 2, $14^{200} + 16^{300}$ has units digit 2.

**Answer:** 2

## Problem 8 Solution

First note 1 has only 1 factor. For a number to have exactly 2 factors, it must be a prime number. The primes from $1 - 30$ are

$$2, 3, 5, 7, 11, 13, 17, 19, 23, 29$$

a total of 10. For a number to have exactly 3 factors, it must be the square of a prime. These are

$$4, 9, 25,$$

a total of 3. The rest of the numbers have $\geq 4$ factors, so there are

$$30 - 1 - 10 - 3 = 16$$

numbers from $1 - 30$ with at least 4 factors.

**Answer:** 16

## Problem 9 Solution

Since $\overline{3a}$ is divisible by 4, the digit $a$ has to be 2 or 6. Using the divisibility rule for 3, the sum of digits $1+2+a+3+a = 6+2a$ is divisible by 3, so $a \neq 2$, leaving $a = 6$. This gives the number 12636. (Double checking, $36 \div 4 = 9$ and $12636 \div 3 = 4212$.)

**Answer:** 12636

## Problem 10 Solution

Note $360° \div 45° = 8$, so the sector is one eighth of a circle with radius 14. The sector there has area

$$\frac{1}{8}\pi \times 14^2 \approx \frac{1}{8} \times \frac{22}{7} \times 14^2 = 77.$$

The triangle is an isosceles right triangle with hypotenuse 14, so it has area

$$\frac{1}{4} \times 14^2 = 49.$$

Hence the shaded region has area

$$L = 77 - 49 = 28.$$

**Answer:** 28

## Problem 11 Solution

If all 24 motorcycles have a capacity of 2 people, the motorcycles will fit

$$24 \times 2 = 48$$

people, meaning that

$$68 - 48 = 20$$

people will be left out. The second kind of motorcycle can hold one extra person, so if we switch 20 of the motorcycles to the second kind everyone will fit. Hence we have $24 - 20 = 4$

motorcycles of the first kind and 20 of the second. Since the first kind costs \$40 per motorcycle and the second kind costs \$30, the total cost to rent the 24 motorcycles is

$$4 \times 40 + 20 \times 30 = 760$$

dollars.

**Answer:** 760

## Problem 12 Solution

Note on the first day Steve has 6 candy bars to choose from. He wants to bring 2 candy bars, and since he is packing them for lunch, the order of the two does not matter. Hence there are

$$\binom{6}{2} = \frac{6 \times 5}{2} = 15$$

choices for the first day. For the second day, there are 4 remaining candy bars, so he has

$$\binom{4}{2} = \frac{4 \times 3}{2} = 6$$

choices. On the last day, there are only 2 remaining, so he only has one choice. Therefore, Steve has

$$15 \times 6 \times 1 = 90$$

total ways to divide the candy bars in 3 groups.

**Answer:** 90

## Problem 13 Solution

$\triangle ABE$ is equilateral, so $AE = BE$ and $\angle EAD = \angle EBC = 90° - 60° = 30°$. Using SAS we know $\triangle AED \cong \triangle BEC$, so $CE = DE$ and thus $\triangle CED$ is isosceles. This implies that $\angle DCE = \angle CDE = 65°$ so

$$\angle CED = 180° - 65° - 65° = 50°.$$

Hence $\angle AEB + \angle CED = 60° + 50° = 110°$, and since $\angle AED = \angle BEC$ we have

$$\angle AED = \angle BEC = \frac{360° - 110°}{2} = 125°.$$

**Answer:** 125

## Problem 14 Solution

Since 5 and 100 are said by the same student, the number of students must be a factor of $100 - 5 = 95$. Among the factors of 95, only 19 is between 10 and 20. So there are 19 students in the class.

**Answer:** 19

## Problem 15 Solution

Jane's parents visit every 6 weeks and grandparents every 8 weeks. They both will visit on weeks that are a common multiple of 6 and 8. Thus they will visit every week that is a multiple of the LCM of 6 and 8, which is 24. Thus both her parents and grandparents will visit Jane during weeks 24 and 48, which is 2 times in 2017.

**Answer:** 2

## Problem 16 Solution

The rocket can land anywhere with 20 meters of the launch site, which is a circle with area

$$\pi \times 20^2 = 400\pi.$$

Candace does NOT need to leave her yard if it lands within 5 meters, which is a circle with area

$$\pi \times 5^2 = 25\pi.$$

Hence the probability Candace needs to leave her yard is

$$1 - \frac{25\pi}{400\pi} = 1 - \frac{1}{16} = \frac{15}{16}.$$

Therefore $P + Q = 15 + 16 = 31$.

**Answer:** 31

### Problem 17 Solution

$B$ is the set of perfect cubes from 1 to 100, so $B$ contains $1^3 = 1$, $2^3 = 8$, $3^3 = 27$, and $4^3 = 64$. Of these, 1 and 64 are perfect squares, so the elements that are in $B$ but not in $A$ are just 8 and 27. Thus the answer is $8 + 27 = 35$.

**Answer:** 35

### Problem 18 Solution

Place 1 ball in each bin, then there are still $20 - 4 = 16$ balls to place in the 4 boxes without restrictions. Using stars and bars we see there are

$$\binom{16 + 4 - 1}{16} = \binom{19}{16} = \binom{19}{3} = 969$$

ways of placing the balls into the boxes.

**Answer:** 969

### Problem 19 Solution

We first find the speed of the train. Since the whole train is on the bridge for 80 seconds, it takes

$$120 - 80 = 40 \text{ seconds}$$

for the train to move on and off the bridge. Since the train takes the same amount of time to move on the bridge and move off the bridge it takes

$$40 \div 2 = 20 \text{ seconds}$$

to do each. If we focus on the front of the train, it takes

$$80 + 20 = 100 \text{ seconds}$$

for the front of the train to travel the 1000 meter length of the bridge. Hence the train moves at a speed of

$$1000 \div 100 = 10 \text{ meters per second.}$$

Knowing the speed we can easily calculate the length. Since it takes 20 seconds to move off the bridge we know the length of the train is

$$20 \times 10 = 200 \text{ meters.}$$

**Answer:** 200

## Problem 20 Solution

Recall the formula for volume of a sphere is $\frac{4}{3}\pi r^3$ and the formula for surface area of a sphere is $4\pi r^2$. Therefore, if the volume of the second ball is 27 times the volume of the first ball, then we see that, since $3^3 = 27$, the radius of the second ball must be $3 \times 4 = 12$ inches. Similarly, since $\left(\frac{1}{4}\right)^2 = \frac{1}{16}$, the radius of the third ball must be $\frac{1}{4} \times 4 = 1$ inches. When stacking the balls, the height will be given by adding all the diameters together. Hence the height is

$$2 \times 4 + 2 \times 12 + 2 \times 1 = 34$$

inches when Billy stacks all 3 balls.

**Answer:** 34

## 2.8  ZIML May 2017 Division M

Below are the solutions from the Division M ZIML Competition held in May 2017.
The problems from the contest are available on p.57.

### Problem 1 Solution
Each of the left and right sides of the long parallelogram have length 2 cm, so each of the top and bottom sides have length

$$(248 - 2 \times 2) \div 2 = 122$$

cm. Each group of 2 rhombi and 2 triangles has top side length 6 cm, so as $122 \div 6 = 20 \, R \, 2$ there are 20 full groups plus one extra rhombus. Hence there are $2 \times 20 + 1 = 41$ total rhombi.

**Answer:** 41

### Problem 2 Solution
Recall a number is divisible by 11 if the alternating sum of its digits is divisible by 11.

View the alternating sum of a five-digit number as two differences plus the final number. For example, the alternating sum of 32232 is $3 - 2 + 2 - 3 + 2 = (3 - 2) + (2 - 3) + 2 = 1 + (-1) + 2 = 2$. The final digit (the units digit) is either 2 or 3. The two differences are each either $0$, $+1$, or $-1$. Hence the the only way to get a multiple of 11 is $(-1) + (-1) + 2 = 0$.

Therefore the five-digit multiple of 11 is 23232 (with alternating sum $2 - 3 + 2 - 3 + 2 = 0$).

**Answer:** 23232

### Problem 3 Solution

On each night all of the $6 + 1 = 7$ people are eligible for the first shift, so there are $\binom{7}{2} = 21$ ways of choosing the first shift. For the second shift, the first two people are not eligible, so there are now $\binom{5}{2} = 10$ ways of choosing who keeps watch. As the second night has the same choices as the first night, there are

$$(21 \times 10)^2 = 210^2 = 44100$$

total ways to decide the 4 shifts.

**Answer:** 44100

**Problem 4 Solution**

There are a total of
$$4 + 5 = 9$$

balls bought. If we tried to save money and bought nine volleyballs instead of four basketballs and five volleyballs, then the price would be
$$4 \times 8 = 32$$

dollars cheaper because each volleyball is $8 less than a basketball. Therefore 9 volleyballs cost

$$185 - 32 = 153$$

dollars in total. Hence we can find the price of one volleyball which is
$$153 \div 9 = 17$$

dollars. Since one basketball is $8 more, a single basketball costs

$$17 + 8 = 25$$

dollars.

**Answer:** 25

## Problem 5 Solution

Let $R$ stand for right and $U$ stand for up. Note that any one of the shortest paths consists of 3 blocks up (3 $U$'s) and 4 blocks right (4 $R$'s). A unique path is determined by the arrangement of 3 $U$'s and 4 $R$'s. The number of possible ways to arrange 3 $U$'s and 4 $R$'s is

$$\binom{7}{3} = \frac{7!}{3! \times 4!} = 35.$$

**Answer:** 35

## Problem 6 Solution

The pattern of the last two digits of powers of 43 is

$$43, 49, 07, 01, 43, 49, 07, 01, \ldots.$$

The cycle length is 4 and $2015 \equiv 3 \pmod{4}$. Thus the last two digits of $43^{2015}$ are given by the third pair of digits in the cycle: 07.

**Answer:** 7

## Problem 7 Solution

Extend lines to make a big rectangle with height 8 and length 16. The shaded region is then the area of the big rectangle, minus a 4 by 4 square and two triangles (with respective bases and heights $16, 4$ and $12, 8$). The area of the shaded region is then

$$16 \times 8 - 4^2 - \frac{1}{2} \times 16 \times 4 - \frac{1}{2} \times 12 \times 8 = 128 - 16 - 32 - 48 = 32.$$

**Answer:** 32

## Problem 8 Solution

At the start, Carolyn has read $\frac{3}{7}$ of the book. After reading 33 additional pages, she has read $\frac{5}{8}$ of the book. Hence, the 33 pages account for

$$\frac{5}{8} - \frac{3}{7} = \frac{11}{56}$$

of the book. Therefore, the book has

$$33 \div \frac{11}{56} = 168$$

pages in total.

**Answer:** 168

## Problem 9 Solution
If we were to list all of the eleven numbers in order, the number right in the middle of the list (the sixth number) would be the average of all of the numbers. Since the sum of the numbers is 374, their average is $374 \div 11 = 34$. Counting backwards by 2 we can see that the third number is $34 - 2 - 2 - 2 = 28$.

**Answer:** 28

## Problem 10 Solution
Since he wants to bring at least 2 each day and at least 4 on Wednesday, he just needs to figure out how to sort the rest of the candy. After saving the minimum amount of candy for each of those days he will have $18 - 2 - 2 - 4 - 2 - 2 = 6$ pieces of candy left. We can use stars and bars to figure out how to arrange those. We can use 6 stars to represent the pieces of candy and $5 - 1 = 4$ bars to represent the days of the week. Therefore, there are

$$\binom{6+5-1}{6} = \binom{10}{6} = \binom{10}{4} = 210$$

possible ways to distribute the rest of the candy throughout the week.

**Answer:** 210

## Problem 11 Solution

The test has

$$12 + 15 + 18 = 45$$

questions in total. For Jong-Zhi to get 75% of them correct, she needs to get

$$75\% \times 45 = 0.75 \times 45 = 33.75$$

questions correct. Thus, to pass Jong-Zhi must get at least 34 questions correct. We know she got 75% of the arithmetic questions correct. Since there are 12 questions, she got

$$75\% \times 12 = 0.75 \times 12 = 9$$

arithmetic questions correct. Similarly, she got 60% of the 15 algebra questions correct which is an additional

$$60\% \times 15 = 9$$

correct questions. Therefore Jong-Zhi has answered

$$9 + 9 = 18$$

questions correct so far. Therefore she needs to answer

$$34 - 18 = 16$$

of the 18 geometry questions to get a passing grade on the test.

**Answer:** 16

## Problem 12 Solution

First note as $\triangle ABC$ is an isosceles right triangle,

$$\angle ABC = (180° - 90°) \div 2 = 45°.$$

Since $\angle CAG = 20°$,

$$\angle AFG = 90° - 20° = 70°.$$

Continuing in this manner, we see

$$\angle CAG = \angle CEG = \angle DEB = 20°$$

and

$$\angle AFG = \angle CFE = \angle DBE = 70°.$$

Therefore,

$$\angle ABD = 70° + 45° = 115°.$$

**Answer:** 115

## Problem 13 Solution

The number of pencils is a multiple of 21, and also a multiple of 28, so it is a multiple of $\text{lcm}(21, 28) = 84$. Thus there are $84k$ pencils, where $k$ is a positive integer. For simplicity we may just assume $k = 1$. Any other value of $k$ also works, and does not affect the result (we may try a different value to see the results). So there are 84 pencils. Based on the description, there are $84 \div 21 = 4$ girls and $84 \div 28 = 3$ boys. So if the teacher distributes the 84 pencils evenly among all 7 students, it will be $84 \div 7 = 12$ pencils for each student.

**Answer:** 12

## Problem 14 Solution

Written in increasing order, the current 9 numbers are

$$1, 1, 2, 2, 3, 5, 7, 8, 9,$$

so the median is 3. Therefore, if the 10th number is at least 4, the median will increase. (For example, if 4 is the 10th number then the new median is $(3 + 4) \div 2 = 3.5$.) As there are 7 numbers that are at least 4 ($4, 5, 6, \ldots, 10$) out of 10 possible choices, the probability is $\dfrac{7}{10}$, so $P + Q = 7 + 10 = 17$.

**Answer:** 17

## Problem 15 Solution

We know that the original ratio of apples to bananas is $3 : 1$. 600 pounds of apples are sold every day. Since

$$3 : 1 = 3 \times 200 : 1 \times 200 = 600 : 200$$

if we assume that only 200 pounds of bananas were sold each day, then the amounts of apples and bananas in the warehouse would always be kept at the ratio $3 : 1$. Then at the end, there were 750 pounds of apples, so there would have been

$$750 \div 3 = 250$$

pounds of bananas remaining. Since each day

$$250 - 200 = 50$$

pounds less of bananas were sold than the actual amount, the number of days was

$$250 \div 50 = 5.$$

Thus at the beginning there were

$$250 \times 5 = 1250$$

pounds of bananas, and

$$1250 \times 3 = 3750$$

pounds of apples.

**Answer:** 3750

## Problem 16 Solution

Recall the volume of a ball is given by $\frac{4}{3} \times \pi R^3$ where $R$ is the radius. Therefore the volume of George's ball is

$$\frac{4}{3} \times \pi \times 6^3 = 288\pi.$$

The combined volumes of Carl's balls are

$$2 \times \left( \frac{4}{3} \times \pi \times 3^3 \right) = 2 \times (36\pi) = 72\pi.$$

Hence the difference is

$$288\pi - 72\pi = 216\pi$$

so $K = 216$.

**Answer:** 216

### Problem 17 Solution

If the $x$-coordinate corresponds to the length of the second stick and the $y$-coordinate corresponds to the length of the third stick, then we can view the lengths of the sticks as being chosen in a $3 \times 5$ rectangle. In this case, the region where the third stick has a longer length than the other two sticks is given by the shaded region (a trapezoid) below

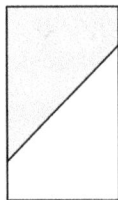

This trapezoid has area half of the full rectangle, so we see that the probability is 50% so $N = 50$.

**Answer:** 50

### Problem 18 Solution

The prime factorization of $252 = 2^2 \times 3^2 \times 7$. To make a perfect cube, we need all the exponents to be a multiple of 3, so if we multiply by $K = 2^1 \times 3^1 \times 7^2$ we get a perfect cube. Therefore $K = 2 \times 3 \times 49 = 294$.

**Answer:** 294

## Problem 19 Solution

We first calculate the distance the bus travels, which is

$$50 \times 18 = 900$$

miles. The truck completes a traveling and resting session every 4 hours. Since

$$18 = 4 \times 4 + 2,$$

in 18 hours the truck will actually be traveling a total of

$$3 \times 4 + 2 = 14$$

hours. Thus, the truck travels

$$42 \times 14 = 588$$

miles. Therefore, the distance between the two starting locations is

$$900 + 588 = 1488$$

miles.

**Answer:** 1488

## Problem 20 Solution

Let $O$ be the center of the circles, and $C$ be the midpoint of $\overline{AB}$.

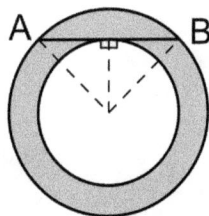

---

Then $\overline{OC} \perp \overline{AB}$, so

$$OA^2 = 1^2 + 1^2 = 2.$$

So the outer circle has radius $\sqrt{2}$. Thus the shaded region has area

$$\pi\sqrt{2}^2 - \pi 1^2 = 2\pi - \pi = \pi \approx 3.14 \approx 3.1$$

so our final answer is 3.1.

**Answer:** 3.1

## 2.9   ZIML June 2017 Division M

Below are the solutions from the Division M ZIML Competition
held in June 2017.
The problems from the contest are available on p.65.

### Problem 1 Solution
We know that a sword costs 9 drachmas more than a spear, so if
Hercules instead buys 7 swords, it will cost

$$304 + 2 \times 9 = 322$$

drachmas. Hence a single sword costs

$$322 \div 7 = 46$$

drachmas.

**Answer:** 46

### Problem 2 Solution
We know that Michelle uses an equal number of pears and oranges.
Since $2 \times 3 = 6$ it will be convenient if we assume she uses 6
pears and 6 oranges. Since 3 pears makes 8 ounces of juice and
Michelle uses
$$2 \times 3 = 6$$
pears, she can make
$$2 \times 8 = 16$$
ounces of pear juice. Similarly, using
$$3 \times 2 = 6$$
oranges, she can make
$$3 \times 8 = 24$$

ounces of orange juice. The total amount of juice is

$$16 + 24 = 40$$

ounces. Since 16 of those ounces are pear juice, the juice blend is

$$\frac{16}{40} = 0.4 = 40\%.$$

pear juice. Therefore, $K = 40$.

**Answer:** 40

## Problem 3 Solution

There are 6 acute angles in total. When adding up all the angles, $\angle 1$ and $\angle 3$ are counted 3 times each, and $\angle 2$ is counted 4 times. We are counting 10 angles in total, and since each of the angles counted is the same, we see that

$$\angle 1 = \angle 2 = \angle 3 = 180° \div 10 = 18°,$$

Hence

$$\angle AOB = 3\angle 1 = 3 \times 18° = 54°.$$

**Answer:** 54

## Problem 4 Solution

Since $36 = 4 \times 9$, we want the number to be divisible by 4 and 9. We use the last two digits for divisibility by 4, so we have that $B = 2$ or $B = 6$ (as 52 and 56 are divisible by 4).

For divisibility by 9 we use the sum of the digits. If $B = 2$ we have $2 + A + 5 + 2 = A + 9$ must be a multiple of 9, so $A = 0$ or $A = 9$, giving the numbers 2052 and 2952. If $B = 6$ we have $2 + A + 5 + 6 = A + 13$ must be a multiple of 9, so $A = 5$.

This gives a third number 2552. The sum of these is therefore $2052 + 2952 + 2556 = 7560$.

**Answer:** 7560

## Problem 5 Solution

We note that the bag contains objects defined by its shape (cubes vs. balls) and color (red vs. blue). Since we are interested in minimizing the number of red balls in the bag, we want to maximize the number of red cubes in the bag.

Given that there is at least one blue cube in the bag, the maximum number of red cubes is $20 - 1 = 19$. Since there are 40 total red objects, the minimum number of red balls is $40 - 19 = 21$. (In this case there will be 21 red balls, 19 red cubes, 9 blue balls, and 1 blue cube.)

**Answer:** 21

## Problem 6 Solution

Note $\triangle ABC$ is an isosceles right triangle, so it is a 45-45-90 triangle. Thus $BC = \sqrt{2}$. As

$$\angle CBD = 180° - 45° = 135°,$$

we have

$$\angle BCD = 180° - 135° - 22.5° = 22.5°,$$

so $\triangle BCD$ is an isosceles triangle. Hence $BD = BC = \sqrt{2}$, so $BD^2 = 2$.

**Answer:** 2

## Problem 7 Solution

Sunday one person is told. Monday 2 people are told. Tuesday $2 \times 2 = 4$ people are told. This pattern continues with 8, 16, 32, and 64 people being told on Wednesday, Thursday, Friday, and Saturday. Hence in total

$$1 + 2 + 4 + 8 + 16 + 32 + 64 = 127$$

people have been told by the end of the following Saturday.

**Answer:** 127

## Problem 8 Solution

Since Iris and Olivia can finish half the project working for 3 and 5 days each and they can complete a third of the project working for 5 and 3 days each, if they each work a total of $3 + 5 = 8$ days, they can complete

$$\frac{1}{2} + \frac{1}{3} = \frac{5}{6}$$

of the project. Therefore they can complete

$$\frac{5}{6} \div 8 = \frac{5}{48}$$

of the project in a single day. Hence, working for 3 days together, Iris and Olivia can finish

$$3 \times \frac{5}{48} = \frac{5}{16}$$

of the project. Since Iris working for 5 days and Olivia for 3 is enough to complete $\frac{1}{3}$ of the work, we see that Iris must be able to finish

$$\frac{1}{3} - \frac{5}{16} = \frac{1}{48}$$

of the project in two days. Hence it will take her

$$2 \div \frac{1}{48} = 96$$

days to complete the entire project by herself.

**Answer:** 96

## Problem 9 Solution

Use Complementary Counting: There are 8! total photographs of all 8 friends in a line. If we group James and Jim together, there are $2! \times (6+1)!$ photographs with James and Jim next to each other. Hence there are

$$8! - 2! \times 7! = 30240$$

photographs with James and Jim separated.

**Answer:** 30240

## Problem 10 Solution

Alice drives an additional 4.5 hours after they meet, so in that time she travels

$$40 \times 4.5 = 180 \text{ miles}$$

to get to site B. It takes Bob

$$180 \div 60 = 3 \text{ hours}$$

to travel these same 180 miles. Since Alice and Bob left at the same time, we know Alice also drove 3 hours from A to the place they met, which makes that distance

$$40 \times 3 = 120 \text{ miles.}$$

So the total distance from A to B is

$$120 + 180 = 300 \text{ miles.}$$

**Answer:** 300

## Problem 11 Solution

Extend the side of the smaller rectangle to cut the bigger rectangle and form a tiny square as shown.

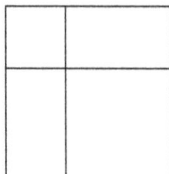

The tiny square has area

$$44 - 28 = 16$$

and hence a side length of 4. The small rectangle thus has dimensions $4 \times 7$ and the lower-right square is 7 by 7 with area $7^2 = 49$.

**Answer:** 49

### Problem 12 Solution

The remainders of the powers of 2 when dividing by 7 follow a pattern:
$$2, 4, 1, 2, 4, 1, \ldots.$$

The length of the cycle is 3. The exponent 2017 is a has a remainder of 1 when divided by 3, so remainder of $2^{2017}$ when divided by 7 is the first number in the cycle: 2.

**Answer:** 2

### Problem 13 Solution

If the school bought only large projectors, they would spend a total of
$$47 \times 700 = 32900$$

dollars, which is

$$32900 - 21200 = 11700$$

dollars over their budget of \$21200. Since there are twice as many medium projectors as small projectors, we can view them as coming in groups of 3, with 1 small and 2 medium projectors. One such group costs a total of

$$1 \times 200 + 2 \times 300 = 800$$

dollars, which is

$$3 \times 700 - 800 = 1300$$

dollars cheaper than a group of 3 large projectors. Hence if the school switches

$$11700 \div 1300 = 9$$

groups of 3 large projectors to groups of 1 small, 2 medium projectors the school will spend the correct amount of money. Thus, the school buys

$$9 \times 2 = 18$$

medium projectors.

**Answer:** 18

### Problem 14 Solution
Calculating the prime factorization of 308000 we get

$$308000 = 2^5 \times 5^3 \times 7 \times 11.$$

To form a factor that is a perfect square, we use even powers of 2, 5, 7, and 11: $2^0, 2^2, 2^4$; $5^0, 5^2$; $7^0$; and $11^0$. As we have a choice for each prime, there are $3 \times 2 \times 1 \times 1 = 6$ combinations which give a perfect square as a factor. (In fact, the perfect square factors are $1, 4, 16, 25, 100, 400$.)

**Answer:** 6

## Problem 15 Solution

Assign the letter $J, P, B, A$ to John, Paul, Ben, and Andy respectively. Since Paul is twice as likely to win, Ben is three times as likely to win, etc., we can think of the winner being chosen randomly from the letters

$$J, P, P, B, B, B, A, A, A, A.$$

(For example, because there are 2 P's, it is twice as likely for Paul to win as needed.) We now have a total of $1 + 2 + 3 + 4 = 10$ letters, with 3 of them the letter B, so the probability Ben wins is

$$\frac{3}{10} = 30\%$$

so $L = 30$.

**Answer:** 30

## Problem 16 Solution

Connect the centers of the circles as shown in the diagram below.

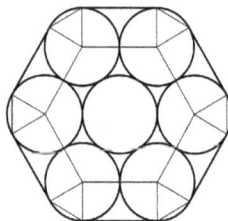

Note this forms a hexagon in the center, 6 rectangles, and therefore 6 equilateral triangles. The rectangles are all 3 mm by 6 mm and the equilateral triangles each have side length 3 mm.

Therefore, the rubber band consists of 6 arcs of 60° each (from circles with radii 3 mm), and 6 line segments of length 6 mm each. The arcs form a full circle, hence the length of the rubber band is

$$2\pi \times 3 + 6 \times 6 = 6\pi + 36 \text{ mm.}$$

Hence $A + B = 6 + 36 = 42$.

**Answer:** 42

### Problem 17 Solution

First note there are $5! = 120$ ways to order the 5 people. Each person needs one of the chairs and we must have one seat between each pair, so that leaves $10 - 5 - 4 = 1$ chair that could either be an extra chair in between two people or on one of the two ends. This means the extra chair can be in one of 6 places. Hence in total there are

$$120 \times 6 = 720$$

different seating arrangements.

**Answer:** 720

### Problem 18 Solution

Since dividing by the integer gives the same remainder, we know the integer must be a factor of

$$571 - 513 = 58 = 2 \times 29,$$

$$658 - 571 = 87 = 3 \times 29,$$

and

$$658 - 513 = 145 = 5 \times 29.$$

The only common factor is 29, so the integer must be 29. As

$$513 \div 29 = 17 \, R \, 20,$$

$$571 \div 29 = 19 \, R \, 20,$$

and

$$658 \div 29 = 22 \, R \, 20,$$

we see the common remainder is 20.

**Answer:** 20

### Problem 19 Solution
Consider the following sample diagram

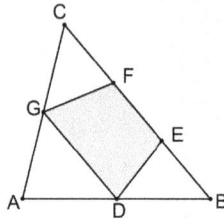

Use $[XYZ]$ to represent the area of $\triangle XYZ$, (so $[ABC] = 36$ is the area of $\triangle ABC$), first note

$$[DEFG] = [ABC] - [DBE] - [FCG] - [ADG].$$

Then, examining triangles that share the same height (but different bases) we have

$$[DBE] = \frac{1}{3}[DBC], \quad [DBC] = \frac{1}{2}[ABC]$$

so

$$[DBE] = \frac{1}{6}[ABC] = 6.$$

Similarly,

$$[FCG] = \frac{1}{6}[ABC] = 6$$

as well. Lastly,

$$[ADG] = \frac{1}{2}[ADC], \quad [ADC] = \frac{1}{2}[ABC]$$

so

$$[ADG] = \frac{1}{4}[ABC] = 9.$$

Thus,
$$[DEFG] = 36 - 6 - 6 - 9 = 15.$$

**Answer:** 15

**Problem 20 Solution**

With the old rule there are

$$26 \times 10^5 = 2600000$$

possible license plates while with the new rule there are

$$26^2 \times 10^4 = 6260000.$$

Hence there would be

$$6260000 - 2600000 = 4160000$$

more possible license plates with the new rule.

**Answer:** 4160000

# 3. Appendix

## 3.1 Division M Topics Covered

**Pre-Algebra and Word Problems***

- Ratios and Proportions: Using ratios to find parts of a whole, Calculating missing information from proportional relationships, Direct and Inverse Proportions, etc.
- Percents: Calculating percent increases and decreases, Relationship between percents and ratios, Using percents in mixture problems (e.g. 40% water and 60% oil)
- Problem Solving Methods: Chicken and Rabbit method, Using ratios when given sums or differences,
- Motion Problems using (Speed)x(Time)=(Distance), Average Speed, Applying direct and inverse proportions to motion problems
- Work using (Rate)x(Time)=(Work Done), Average Rate of Work, Applying direct and inverse proportions to work problems

*Note: Setting up and solving equations is not necessary for any of the problems in Division M. Students are allowed to use equations to solve the questions, but the questions are designed

---

to be solved without using equations or systems of equations.

## Geometry

- Areas and Perimeters of Basic Shapes such as triangles, rectangles, parallelograms, trapezoids, and circles
- Angles in Parallel Lines (corresponding angles, alternating interior/exterior angles, same-side interior/exterior angles, etc.)
- Triangles: Congruence and Similarity, Pythagorean theorem, Ratios of Sides for triangles with angles of $45°, 45°, 90°$ or $30°, 60°, 90°$
- Interior and Exterior Angles of Polygons, including the sum of all the interior or exterior angles, the measure of each angle if the polygon is regular, etc.
- Geometric Reasoning with Areas: Congruent shapes have the same area, Similar triangles have a ratio of areas that is the square of the ratio of their sides, Triangles with the same height have a ratio of their areas equal to the ratio of their bases, etc.
- Circles: Arc Length, Sector Area, Definitions for Tangent Lines and Tangent Circles
- Volumes and Surface Areas of Basic Solids such as cubes, spheres, rectangular prisms (boxes), and pyramids

## Counting and Probability

- Sum and Product Rules
- Permutations and Combinations
- Counting Methods: Complementary counting, Stars and bars (also called sticks and stones, balls and urns, etc.), Grouping objects that must be together, Inserting objects that must be apart into spaces between objects, etc.
- Sequences: Arithmetic and Geometric Sequences, Sum of elements in an arithmetic sequence, Finding patterns for general

sequences
- Probability and Sets: Definitions for event, sample space, complement, intersection, and union, Understanding the use of Venn Diagrams
- Probability in Finite Sample Spaces: Probability as a ratio of outcomes, Probabilities sum to 1, Computing probabilities with complements
- Geometrical Probability: Probability as a ratio of lengths, areas, or volumes
- Basic Statistics: Mean (Average), Median, Mode for lists, Interpreting data from graphs, bar charts, tables, etc.

**Number Theory**

- Fundamental Definitions: Prime numbers, factors/divisors, multiples, least common multiple (LCM), greatest common factor/divisor (GCF or GCD), perfect squares/cubes/etc.
- Divisibility Rules for numbers such as 2, 3, 4, 5, 8, 9, 10, 11, and how to combine the rules for numbers such as 6, 22, etc.
- (Unique) Prime Factorization and using the prime factorization to find the number of factors, to test whether a number is a perfect square/cube/etc, to find the LCM or GCD, etc.
- Factoring Tricks: Factors come in pairs, perfect squares have an odd number of factors, etc.
- Remainders and Patterns: Finding the units digit, finding the last two digits, finding the remainder when divided by 11, etc.
- Basic Modular Arithmetic: Understand "congruent modulo $m$" means two numbers have the same remainder when divided by $m$, The sum of two numbers is congruent modulo m to the sum of the remainders of the numbers when divided by $m$

## 3.2  Glossary of Common Math Terms

**Acute Angle**  An angle less than $90°$.

**Altitude of a Triangle**  A line segment connecting a vertex of a triangle to the opposite side forming a right angle. Also called the height of a triangle.

**Angle**  A figure formed by two rays sharing a common vertex. Often measured in degrees.

**Arc**  The curve of a circle connecting two points.

**Area**  The amount of space a region takes up. Often denoted using square brackets: area of $\triangle ABC = [ABC]$.

**Arithmetic Sequence**  A sequence where the difference between one term and the next is constant.

**Average**  See Mean.

**Base of a Triangle**  One side of a triangle, often used when the altitude is drawn from the opposite side to this base.

**Binomial Coefficient**  The symbol $\binom{n}{k} = \dfrac{n!}{k!(n-k)!}$.

**Chord**  A line segment connecting two points on the outside of a circle.

**Circle**  A round shape consisting of points that all have the same distance (called the radius) from the center of the circle.

**Circumference**  The perimeter of a circle.

**Composite Number**  A number that is not prime.

**Congruent** Two shapes or figures that are exactly the same.

**Cube** A solid figure formed by 6 congruent squares that all meet at right angles.

**Deck of Cards** A standard deck of cards has 52 cards. There are 4 suits (clubs, diamonds, hearts, and spades) with each suit having cards of 13 ranks ($A$ (ace), $2, 3, \ldots, 10$, $J$ (jack), $Q$ (queen), and $K$ (king) ).

**Denominator** The bottom number in a fraction.

**Diagonal** A line segment connecting two vertices of a shape or solid that is not an edge of the shape or solid.

**Diameter** A chord passing through the center of a circle. The diameter has length that is twice the radius.

**Die or Dice** A standard die (plural is dice) has 6 sides. Each of the 6 sides has the same chance when the die is rolled.

**Digit** One of $0, 1, 2, \ldots, 9$ used when writing a number.

**Distinguishable Objects** Objects that are different.

**Divisible** A number is divisible by another number if there is no remainder when the first number is divided by the second. For example, 35 is divisible by 7.

**Divisor** A number that evenly divides another number. For example, 6 is a divisor of 48. Also called a factor.

**Edge** A line segment connecting two vertices on the outside of a shape or solid.

**Equally Likely** Having the same chance of occurring.

**Equiangular Polygon** A shape with all equal angles.

**Equilateral Polygon** A shape with all equal sides.

**Equilateral Triangle** A regular triangle, one with three equal sides and three equal angles.

**Even Number** A number divisible by 2.

**Exponent** The number another number is raised to for powers. For example, in $a$ to the power of $b$ ($a^b$), the exponent is $b$.

**Face** The shape or polygon on the outside of a solid region.

**Factor of a Number** A number that evenly divides another number. For example, 6 is a factor of 48. Also called a divisor.

**Factorial** The symbol ! where $n! = n \times (n-1) \times (n-2) \cdots \times 1$.

**Fraction** An expression of a quotient. For example, $\frac{1}{2}$ or $\frac{9}{7}$.

**Geometric Sequence** A sequence where the ratio between one term and the next is constant.

**Greatest Common Divisor/Factor (GCD/GCF)** The largest number that is a divisor/factor of two or more numbers.

**Indistinguishable Objects** Objects that are the same.

**Intersecting** Lines or curves that cross each other.

**Intersection of Two Sets** The set of objects that are in both of the two sets. Denoted using $\cap$. For example, $\{2,3\} \cap \{3,4,5\} = \{3\}$.

**Isosceles Triangle** A triangle with two equal sides and two equal angles.

**Least Common Multiple (LCM)** The smallest number that is a multiple of two or more numbers.

**Mean** The sum of the numbers in a list divided by the how many numbers occur in the list. Also called the average.

**Median** The number in the middle of a list when the list is arranged in increasing order.

**Midpoint** The point in the middle of a line segment.

**Mode** The number or numbers occurring most often in a list of numbers.

**Multiple** A number that is an integer times another number. For example, 72 is a multiple of 8.

**Numerator** The top number in a fraction.

**Obtuse Angle** An angle between $90°$ and $180°$.

**Odd Number** A number not divisible by 2.

**Parallel Lines** Lines that do not intersect.

**Perfect Cube** A number that is another number cubed. For example, $64 = 4^3$ is a perfect cube.

**Perfect Square** A number that is another number squared. For example, $64 = 8^2$ is a perfect square.

**Perimeter** The length/distance around the outside of a shape.

**Pi ($\pi$)** A number used often in geometry. $\pi = 3.1415926\ldots \approx$ $3.14 \approx \dfrac{22}{7}$.

**Polygon** A shape formed by connected line segments.

**Prime Factorization** The expression of a number as the product of all its prime factors. For example, 24 has prime factorization $2 \times 2 \times 2 \times 3 = 2^3 \times 3$.

**Prime Number** A number whose only factors are one and itself.

**Proportional Ratios** Ratios that have equal values when expressed in fraction form. For example, $2:3$ is proportional to $8:12$.

**Quadrilateral** A shape with four sides.

**Quotient** The integer quantity when dividing one number by another. For example, the quotient of $38 \div 5$ is 7 as $38 = 7 \times 5 + 3$.

**Radius of a Circle** The distance from the center of the circle to any point on the outside of the circle.

**Randomly** Chosen for a group of objects. Unless specified, the chance of choosing each object is the same as any other object.

**Rank of a Card** See Deck of Cards.

**Ratio** A relation depicting the relation between two quantities. For example $2:3$ or $\dfrac{2}{3}$ denotes that for every 3 of the second quantity there are 2 of the first quantity.

**Rational Number** A number that can be written as a fraction.

**Reciprocal** One divided by the number. For example, the reciprocal of 7 is $\frac{1}{7}$.

**Rectangle** A quadrilateral with four right angles (an equiangular quadrilateral).

**Regular Polygon** A polygon with all equal sides and all equal angles (equilateral and equiangular).

**Remainder** The quantity left over when one integer is divided by another. For example, the remainder of $38 \div 5$ is 3 as $38 = 7 \times 5 + 3$.

**Rhombus** A quadrilateral with four equal sides (an equilateral quadrilateral).

**Right Angle** A $90°$ angle.

**Right Triangle** A triangle containing a right angle.

**Scalene Triangle** A triangle with three unequal sides and three unequal angles.

**Sector** The region formed by an arc and the two radii connecting the ends of the arc to the center of the circle.

**Sequence** An ordered list of numbers.

**Set** An unordered collection or group of objects without repeated elements. Denoted using curly brackets. For example, $\{1,2,3,4\}$ is the set containing the integers $1,\ldots,4$.

**Similar** Shapes or solids that have the same angles and sides that share a common ratio.

**Simplest Radical Form** An expression containing a radical such that the number inside the radical is an integer that has no perfect squares.

**Sphere** A round solid consisting of points that all have the same distance (called the radius) from the center of the sphere.

**Square** A shape with four equal sides and four equal angles (a regular quadrilateral).

**Subset** A set of objects that is contained inside a larger set of objects. Denoted using $\subseteq$. For example $\{2,3\} \subseteq \{1,2,3,4\}$.

**Suit of a Card** See Deck of Cards.

**Surface Area** The total area of all the faces of a solid.

**Trapezoid** A quadrilateral with one pair of parallel sides.

**Triangle** A shape with three sides.

**Union of Two Sets** The set of objects that are in one or both of the two sets. Denoted using $\cup$. For example, $\{2,3\} \cup \{3,4,5\} = \{2,3,4,5\}$.

**Venn Diagram** A diagram with circles used to understand the relationship between overlapping sets.

**Vertex** The intersection of line segments, especially the intersection of sides or edges in a shape or solid.

**Volume** The amount of space a solid region takes up.

**With Replacement** When choosing objects with replacement, a chosen object is returned to the others allowing it to be chosen more than once.

ZOOM INTERNATIONAL MATH LEAGUE: ziml.areteem.org

## .3 ZIML Answers

## ZIML October 2016 Division M

| | | | |
|---|---|---|---|
| Problem 1: | 10 | Problem 11: | 3 |
| Problem 2: | 75 | Problem 12: | 0 |
| Problem 3: | 13 | Problem 13: | 80 |
| Problem 4: | 5 | Problem 14: | 64 |
| Problem 5: | 240 | Problem 15: | 78 |
| Problem 6: | 30 | Problem 16: | 115 |
| Problem 7: | 35424 | Problem 17: | 60 |
| Problem 8: | 121 | Problem 18: | 3 |
| Problem 9: | 2184 | Problem 19: | 8 |
| Problem 10: | 1 | Problem 20: | 350 |

# ZIML November 2016 Division M

Problem 1:   10

Problem 2:   16

Problem 3:   12

Problem 4:   8

Problem 5:   165

Problem 6:   18

Problem 7:   90

Problem 8:   760

Problem 9:   70

Problem 10:  21

Problem 11:  13

Problem 12:  48

Problem 13:  4

Problem 14:  6

Problem 15:  33

Problem 16:  100

Problem 17:  8

Problem 18:  75

Problem 19:  9

Problem 20:  45

# ZIML December 2016 Division M

Problem 1:   2000          Problem 11:  6

Problem 2:   44            Problem 12:  21

Problem 3:   44            Problem 13:  0

Problem 4:   3             Problem 14:  20

Problem 5:   1045          Problem 15:  16

Problem 6:   20            Problem 16:  12

Problem 7:   80            Problem 17:  135

Problem 8:   24            Problem 18:  46

Problem 9:   50            Problem 19:  60

Problem 10:  4             Problem 20:  217

# ZIML January 2017 Division M

| | | | |
|---|---|---|---|
| Problem 1: | 25 | Problem 11: | 18 |
| Problem 2: | 49 | Problem 12: | 15 |
| Problem 3: | 3555 | Problem 13: | 13 |
| Problem 4: | 15 | Problem 14: | 35 |
| Problem 5: | 720 | Problem 15: | 1440 |
| Problem 6: | 54 | Problem 16: | 15 |
| Problem 7: | 252 | Problem 17: | 886 |
| Problem 8: | 8 | Problem 18: | 20 |
| Problem 9: | 252 | Problem 19: | 60 |
| Problem 10: | 410 | Problem 20: | 17 |

# ZIML February 2017 Division M

Problem 1:   5                  Problem 11:   90

Problem 2:   15                 Problem 12:   21

Problem 3:   40                 Problem 13:   840

Problem 4:   45                 Problem 14:   1526

Problem 5:   39                 Problem 15:   56

Problem 6:   10                 Problem 16:   12

Problem 7:   90                 Problem 17:   35828

Problem 8:   36                 Problem 18:   76

Problem 9:   36                 Problem 19:   2

Problem 10:  8                  Problem 20:   61

## ZIML March 2017 Division M

| | | | |
|---|---|---|---|
| Problem 1: | 15 | Problem 11: | 7 |
| Problem 2: | 56 | Problem 12: | 2048 |
| Problem 3: | 12 | Problem 13: | 48 |
| Problem 4: | 57 | Problem 14: | 13 |
| Problem 5: | 70 | Problem 15: | 17 |
| Problem 6: | 98 | Problem 16: | 3 |
| Problem 7: | 5 | Problem 17: | 150 |
| Problem 8: | 7 | Problem 18: | 4 |
| Problem 9: | 75 | Problem 19: | 6 |
| Problem 10: | 2 | Problem 20: | 36 |

# ZIML April 2017 Division M

**Problem 1:**  12

**Problem 2:**  12.5

**Problem 3:**  109

**Problem 4:**  36

**Problem 5:**  15

**Problem 6:**  25

**Problem 7:**  2

**Problem 8:**  16

**Problem 9:**  12636

**Problem 10:**  28

**Problem 11:**  760

**Problem 12:**  90

**Problem 13:**  125

**Problem 14:**  19

**Problem 15:**  2

**Problem 16:**  31

**Problem 17:**  35

**Problem 18:**  969

**Problem 19:**  200

**Problem 20:**  34

## ZIML May 2017 Division M

| | | | |
|---|---|---|---|
| Problem 1: | 41 | Problem 11: | 16 |
| Problem 2: | 23232 | Problem 12: | 115 |
| Problem 3: | 44100 | Problem 13: | 12 |
| Problem 4: | 25 | Problem 14: | 17 |
| Problem 5: | 35 | Problem 15: | 3750 |
| Problem 6: | 7 | Problem 16: | 216 |
| Problem 7: | 32 | Problem 17: | 50 |
| Problem 8: | 168 | Problem 18: | 294 |
| Problem 9: | 28 | Problem 19: | 1488 |
| Problem 10: | 210 | Problem 20: | 3.1 |

# ZIML June 2017 Division M

| | | | |
|---|---|---|---|
| **Problem 1:** | 46 | **Problem 11:** | 49 |
| **Problem 2:** | 40 | **Problem 12:** | 2 |
| **Problem 3:** | 54 | **Problem 13:** | 18 |
| **Problem 4:** | 7560 | **Problem 14:** | 6 |
| **Problem 5:** | 21 | **Problem 15:** | 30 |
| **Problem 6:** | 2 | **Problem 16:** | 42 |
| **Problem 7:** | 127 | **Problem 17:** | 720 |
| **Problem 8:** | 96 | **Problem 18:** | 20 |
| **Problem 9:** | 30240 | **Problem 19:** | 15 |
| **Problem 10:** | 300 | **Problem 20:** | 4160000 |

www.ingramcontent.com/pod-product-compliance
Lightning Source LLC
Chambersburg PA
CBHW050117210326
41519CB00015BA/3999